T0342516

NEXT-GENERATION
VIDEO CODING
AND STREAMING

NEXT-GENERATION VIDEO CODING AND STREAMING

BENNY BING

WILEY

For general information on our other products and services or for technical support, please contact our Customer Care Department within the United States at (800) 762-2974, outside the United States at (317) 572-3993 or fax (317) 572-4002.

Wiley also publishes its books in a variety of electronic formats. Some content that appears in print may not be available in electronic formats. For more information about Wiley products, visit our web site at www.wiley.com.

Library of Congress Cataloging-in-Publication Data:

Bing, Benny.
  Next-generation video coding and streaming / Benny Bing.
    pages cm
  Includes bibliographical references and index.
  ISBN 978-1-118-89130-8 (hardback)
1. Video compression. I. Title.
  TA1638.B56 2015
  006.6'96–dc23

2015020396

Cover image courtesy of Godruma/Getty

10 9 8 7 6 5 4 3 2 1

1 2015

CONTENTS

PREFACE

TV remains the single most important and engaging source of information and entertainment. U.S. teenagers spend more than three times of their spare time watching TV than on social media. The global footprint of TV has been enhanced recently by online video, which includes online TV. U.S. consumers watch more movies online than on DVDs, Blu-ray discs and other physical video formats. This trend is driven by the flexibility of on-the-go mobile entertainment and the widespread adoption of video-capable smartphones and tablets. These personal devices have become ubiquitous with greatly expanded computing power and memory, improved displays, and network connectivity. The accelerated growth of video traffic on the Internet is expected to continue. However, supporting high-quality video delivery presents a significant challenge to Internet service providers due to the higher bandwidth demands compared to data and voice traffic.

This book describes next-generation video coding and streaming technologies with a comparative assessment of the strengths and weaknesses. Specific emphasis is placed on the H.265/HEVC video coding standard and adaptive bit rate video streaming. H.265/HEVC has been developed to meet the demands of emerging UHD video services and pervasive online video streaming. The commercial adoption of H.265/HEVC has started to gain traction since 2014. Invaluable insights into the coding efficiencies of the intracoded and intercoded frames are described in this book, including the impact of different types of video content and powerful feature sets such as the hierarchical block structure and new coding parameters. Adaptive streaming is a key enabling technology that can achieve smooth and reliable video delivery over heterogeneous wireline and wireless networks, as well as multiscreen personal devices. It provides autonomous bandwidth management and maintains quality of service even as link conditions and network congestion vary. This book provides an in-depth study on the practical performance of the popular adaptive streaming platforms and useful tips for streaming optimization. Innovative techniques related to aggregate adaptive stream bandwidth prediction, duplicate chunk suppression, and server-based adaptive streaming are also discussed.

I wish to thank Wiley's Publisher Dr. Simone Taylor, for her encouragement and patience in overseeing this book project. I also like to acknowledge my industry collaborators and former students who have been generous in sharing many useful comments. The book includes over 220 illustrative figures and over 110 homework problems containing interesting ideas and extensions to key concepts. Powerpoint slides and solutions to the homework problems are available to instructors who adopt the book for a course. Please feel free to send your comments and questions to bennybing@yahoo.com.

BENNY BING

1

DIGITAL VIDEO DELIVERY

Television has in many ways promoted understanding and cooperation among people all over the world. About 600 million people saw the first person walk on the moon and a billion people watched the 20th Summer Olympic Games. By 2012, there were over 3.6 billion viewers for the 30th Summer Olympic Games. TV watching used to be confined primarily to the living room. This has changed. The ubiquity of HD-capable smartphones and tablets equipped with powerful video decoders enables TV view time on mobile devices to surpass view time on the TV. The Internet has become a key media distribution platform that has opened up new ways for discovering, sharing, and consuming TV content anywhere, anytime, and on any device. Online Internet TV providers are trumping cable and satellite pay-TV providers with a dramatic increase in subscription and advertising revenue in recent years. Ala-carte-style Internet TV has now started to break the traditional pay-TV distribution model that is based on channel bundling. Although pay-TV providers are making TV content available online for their subscribers, they may now have to migrate to online streaming boxes and ditch the venerable set-top. Cable is now a broadband business. Among the top cable providers, broadband Internet service accounted for more subscribers than cable TV. Ultimately, pay-TV providers may have to rely on broadband to grow profits. Currently, over 10 million US households are broadband-only. In this chapter, we analyze these game-changing trends in digital video delivery.

Next-Generation Video Coding and Streaming, First Edition. Benny Bing.
© 2015 John Wiley & Sons, Inc. Published 2015 by John Wiley & Sons, Inc.

1.1 BROADBAND TV LANDSCAPE

Over 80% of Internet users watch video while 30% of these users watch TV. In the United States, viewers spent an average of more than 6 h/month watching video on the Internet. Streaming live sports programs online makes truly national or global events possible, reaching millions of consumers via handheld devices. There are several challenges. Due to the mobility of subscribers and the heterogeneity of the user devices, the streaming server has to adapt the video content to the characteristics and limitations of both the underlying network and the end devices. These include variations in the available network bandwidth and user device limitations in processing power, memory, display size, battery life, or download limits.

1.1.1 Internet TV Providers

The emergence of over-the-top (OTT) online content providers such as Netflix, Hulu, and Amazon offers more choices to the consumer by providing replacement or supplementary TV services, usually TV shows and movies but no sports programming. The service is either free or much cheaper than pay-TV and this has led to a steady migration of subscribers from pay-TV to online TV, despite efforts from cable and satellite pay-TV providers in making TV content available online. Unlike OTT providers, these "TV Everywhere" Web portals may include both sports and video on demand (VOD) or time-shifted TV content.

Because broadcast pay-TV tends to surpass online TV in visual quality, it delivers better overall experience. For example, the quality of online TV service may fluctuate according to the bandwidth availability on the broadband Internet connection. However, other factors such as choice of content, flexible viewing time, and content portability are also important for the consumer. Thus, hardware set-top box (STB) vendors (e.g., TiVO) traditionally aligned to linear broadcast programming cable TV service have integrated OTT streaming content to their channel lineup.

Pay-TV operators may follow this lead and let OTT content into their STB. In doing so, subscription rates may be lower compared to traditional pay-TV subscription. For example, Walt Disney recently signed a carriage deal with Dish Network, making it the first pay-TV provider to bundle ABC, ESPN, and other channels owned by Disney in a TV service delivered entirely over the Internet. This lower-cost product will allow Dish to broaden their customer base and target new broadband-only consumers who do not currently subscribe to any form of cable or satellite TV. It is interesting to note that Dish is employing a small-scale version of the traditional multichannel subscription bundling, which provides carriage fees to the TV industry for large packages of channels.

Ala-carte-style Internet TV, where users can subscribe to individual channels, has now arrived. The new age of Web-delivered TV allows viewers to have more options to pay only for the TV networks or programs they want to watch and to decide how, when, and where to watch them. Unlike pay-TV, many of these subscription-based video on demand (SVOD) providers are currently ad-free. However, ad-based OTT service may appear in future to further reduce subscription fees. As live and on-demand Internet TV programming becomes mainstream, this development will ultimately increase competition and further drive subscription prices down. It has already forced some of the biggest pay-TV providers in the United States to merge.

1.1.2 Netflix

Netflix is currently the leader of OTT providers. It is a SVOD service where regular subscribers pay a low rate of $7.99 per month and ultra-high definition (UHD) customers pay $11.99 per month. There are over 80 million Netflix-capable devices, including TVs, smartphones, tablets, and game consoles. The number of Netflix viewers passed the number of YouTube viewers. Netflix has over 57 million subscribers in 50 countries (about 39 million US subscribers and 18 million foreign subscribers) and streams over 2 billion hours of TV shows and movies per month. Roughly half of all US households now have a Netflix subscription. Netflix accounts for nearly 30% of Web traffic in the United States at peak periods, a dominant leader among all online video websites. This percentage has increased as Netflix has added 4K UHD content to its streaming video library. Such data-heavy usage is creating a huge problem for Internet service providers (ISPs), who are demanding higher fees for the interconnection required to deliver high-quality service.

1.1.3 Hulu

Hulu provides both free TV and SVOD services. It handles over 30 million online users (over 6 million are paid subscribers) and over 1 billion video streams per month. Hulu and Amazon account for 1–2% of all Web traffic during peak hours.

1.1.4 Amazon

Amazon Prime Instant Video has a few million subscribers. Six new original TV series have been launched by Amazon in 2014, including five programs that were produced in UHD format. Amazon Studios also plans to shoot its new drama and comedy series pilots in UHD, teaming up with Samsung and major media corporations including Warner Bros and Lionsgate.

1.1.5 YouTube

YouTube has the largest library of both user-generated and premium videos. The growth of YouTube is accelerating in spite of increased competition from social networks such as Facebook. Unlike Netflix, which offers full-length movies and TV shows, YouTube's short-form videos are particularly popular. These short-duration videos are perfect for on-the-go viewing on small-screen personal devices such as smartphones. Thus, YouTube dominates other online TV websites with over 20% of all mobile downstream traffic in the Unites States. Roughly the same amount of traffic is delivered over fixed networks during peak hours. This is about half of Netflix, even though YouTube has a far greater number of views and downloads. For example, in May 2011, the number of views on YouTube hit 3 billion/day. The first video posted on YouTube was a 19-s clip called *Me at the Zoo* over 10 years ago. Today, more than 300 h of video are uploaded every minute. Google's Hangouts enable virtual participation in live events where users may record and stream videos, as well as interact in conferences, music concerts, and even football matches. The free service works on any Android and Apple device and recorded videos can be broadcast on YouTube.

1.1.6 ESPN3

Unlike OTT providers, ESPN3 offers live sports viewing online. The 2012 Super Bowl attracted over 2.1 million unique viewers when the game was streamed online by ESPN3 in the United States for the first time. In that year, all 302 events of the summer Olympics were streamed live. The service is available to Internet or pay-TV subscribers from affiliated service providers who pay fees to ESPN. Since 2008, free ESPN3 service has been made available to US college/university students and military personnel.

1.1.7 HBO

HBO has been one of pay-TV's most successful products for decades but plans to break off from the cable bundle in 2015 and distribute its shows to consumers using a standalone OTT streaming service via Apple TV. It will become a direct competitor to Netflix's SVOD service.

1.1.8 CBS

CBS launched a new subscription Internet TV streaming service on October 2014 that allows people to watch its live programming and thousands of its current and past shows on demand without paying for a traditional TV subscription. The new "CBS All Access" service costs $5.99 a month. CBS is the first traditional broadcaster that makes a near-continuous live feed of its local stations available over the Internet to non-pay-TV subscribers.

1.1.9 Sony

Sony launched the world's first "Video Unlimited" UHD movie/TV streaming service on September 4, 2013. Sony has Internet rights to carry channels from Viacom, which owns cable channels such as MTV, Nickelodeon, and Comedy Central. Sony is also developing an original TV drama series that will be available initially on its PlayStation gaming consoles.

1.1.10 Retail Giants

Retail giants such as Best Buy, Sears, and Walmart are joining the digital media ecosystem. Best Buy, for example, is providing their high-dollar customers with free online video rentals from CinemaNow.

1.2 INTERNET TV DELIVERY PLATFORMS

Content distribution platforms to store, transcode, and deliver petabytes of video on commodity hardware are readily available. The TV platforms are scalable in computing, storage, and bandwidth resources even when the data centers are interconnected through the Internet and long-distance networks. The dynamic location of these service facilities and the virtualization of hardware and software elements are increasingly complemented by a flexible software-defined network (SDN) architecture, which employs programmable application interfaces to couple the control and data planes of network hardware (e.g., routers, switches) so that data can be pulled and resources can be reconfigured from any connected device on the network.

1.2.1 Cloud TV

Cloud TV leverages on the well-known concept of cloud computing to provide video storage and streaming services for content and pay-TV providers. Cloud computing is a common metaphor for computation as well as data storage, access, and management, freeing the end user from worries about the location and management of the resources they use. Some examples of cloud computing platforms include Amazon Web Services that supports companies such as Netflix and Pinterest, and Google Cloud Platform that enables developers to build and deploy applications on Google's cloud infrastructure. Big media and entertainment companies are becoming more comfortable with cloud technologies and cloud storage solutions. By putting as much business operation as possible in the cloud, service providers can operate far more efficiently with reduced capital and operating expenses due to hands-off maintenance and customers may benefit. For example, by shifting the DVR from on-premise infrastructure to the cloud, video recordings can be scheduled and viewed from anywhere. Cloud networks are complemented by file acceleration technologies that allow large files to be transmitted to cloud processing locations quickly and securely.

Cloud TV promises ubiquitous access and security without the administrative chores of managing hardware, upgrades, and backups. Scalability issues, server bottlenecks, and server failures are alleviated to a great extent. This is achieved by providing content storage and streaming on high-end servers from strategically located data centers. The emergence of cloud TV offerings will have a significant impact on both fixed and mobile communication networks. Stronger interworking and interoperability between system and network elements are needed because data centers are becoming containers of virtual provider networks. Efficient resource management in geographically distributed data centers and cloud networks to provide bandwidth guarantees and performance isolation is particularly important with the increasing reliance on bandwidth-demanding virtual machine migrations for resource consolidation.

Cloud TV is integral to enabling the flourishing electronic sell-through business that pay-TV distributors have launched, which allow consumers to purchase permanent copies of movies through cable VOD platforms. For example, Warner Bros, NBC Universal, and other studios support the UltraViolet platform, which allows consumers who purchase DVDs in retail outlets to access the movies online. More recently, Disney Movies Anywhere launched a cloud-based movie service that allows viewers who buy movies for viewing on Apple devices to view the content online. Disney Movies Anywhere has 400 titles from Disney, Pixar, and Marvel libraries.

1.2.2 Content Delivery Network

Cloud service providers normally employ a large number of servers in one location. In contrast, content delivery or distribution networks (content delivery network, CDNs) allow the servers to be distributed in different locations and works on the principle of delivering content from a server (an edge server) that is nearest to the user's location. CDNs are widely deployed in electronic commerce, media content delivery and caching, software distribution, and multiplayer immersive games. CDNs are necessary to publish rich media, including HD video and audio, because traditional Web hosts serve media from only one location, running the risks of network congestion on the public Internet. Moreover, the performance of traditional Web hosts degrades significantly as the number of simultaneous remote connections increase. Thus, there are no predictable network guarantees because the solution cannot scale globally. The CDN replicates the stream from the source at its origin server and caches the stream in many or all of its edge servers. The user that requests the stream is redirected to the closest edge server. Thus, regardless of device and location, users will experience better performance with CDNs because the media will be automatically delivered to them from a server in their location and very often, one directly connected to their ISPs.

Super points of presence (POPs) or broadcast nodes are normally built in major cities around the world, which allows the public Internet to be bypassed, thereby improving media delivery. For example, media startup times can be minimized and higher quality media streams can be supported because the shorter network span between the edge server and client reduces the number of clients accessing the server, hence increasing the available bandwidth. Moreover, caching of static content can further improve performance whereas dynamic elements of a specific application can be tuned and delivered using an application delivery network. CDN security services prevent content theft and provide access control. Many CDNs employ HTTP servers. HTTP allows standard caching (which can scale to a large number of connections) and eliminates any issues with firewalls since it uses the same ports as Web browsing. By not using proprietary video streaming protocols or specialized streaming servers, they can avoid additional capital expenditure. CDN providers are offering management software with greater control to enterprises and operators.

1.2.3 Free CDN

Netflix's Open Connect [1] is a CDN that allows ISPs to connect directly for free. ISPs can do this either by free peering with Netflix at common Internet exchanges or placing Netflix's free storage appliances in or near their network, thereby saving additional transit costs. Free peering allows voluntary interconnection of administratively

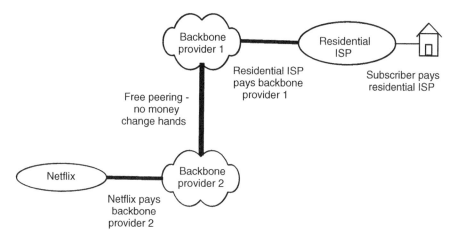

Figure 1.1 Free peering.

separate backbone Internet networks, with no money changing hands between backbone providers (Figure 1.1). Netflix also provides details of hardware design and the open source software components of the server. The Open Connect appliance hardware provides very high storage density (100 TB) and high throughput (10 Gbit/s via an optical network connection).

1.2.4 Video Transcoding

Video transcoding is equivalent to video streaming except that the output is sent to a file instead. It is common in some deployments to maintain 15 different versions of the same movie or TV program. Transcoding selects the appropriate codec and bit rate for the delivery network, a key enabler for anywhere, anytime, any device TV delivery. Software for transcoding has matured and now produces a broad range of output formats from Web and mobile standards such as MPEG-4, Windows Media, QuickTime, and Flash.

1.3 SECOND SCREEN DEVICE ADOPTION

While TVs are the traditional first screen of many residential homes, watching TV shows on second screen devices such as phones or tablets has become popular. Global smartphones sales topped one billion for the first time in 2013. Apple has sold over 1 billion iOS devices by early 2015, doubling the number 2 years ago. Currently, 10 iPhones are selling every second. According to the *New York Times* (September 29, 2013), Apple passes Coca-Cola as the most valuable brand. The rising demand for tablets is driven not only by young adults and working professionals but also kids. Many parents are buying inexpensive mini-tablets for their children. Tablets are increasingly being used in schools because the touch-based screens are a natural tool for teaching and learning. This development is far more successful than initiatives (launched several years ago) to build affordable laptops for kids.

1.3.1 Mobile Video

Video traffic constitutes a major percentage of mobile traffic, more than twice the volume of data and voice traffic. Smartphones and tablets typically allow only wireless connectivity and do not come with wired network interfaces. The proliferation of these personal devices led to pervasive mobile access. Unlike laptops, many smartphones and tablets also come with high-definition multimedia interface (HDMI) connectivity. TV programs on tablets and smartphones can also be streamed wirelessly (via Wi-Fi) to a TV using a Wi-Fi HDMI dongle that is plugged into the TV set. In essence, phones and tablets have become a DVR, portable entertainment player, TV remote, and media server conveniently integrated into one handheld device.

Verizon supports wireless OTT services and currently offers 20 TV program channels to mobile users. It expects to deliver over 100 channels and will leverage on multicast technology to enable efficient 24-h TV programming on the wireless network. Sports, news, concerts, and other live content can be viewed in real-time. An ecosystem can be created around such content that does not affect the traditional linear TV model. The mobile video service is complemented by Verizon's migration to all-IP video servers, which removes the need for STBs. This will benefit customers with faster residential service access using one piece of equipment in the home.

1.3.2 Mobile Versus Traditional TV

Although view time on mobile devices has now surpassed view time on TV, mobile TV is not a substitute for traditional TV. The most obvious difference is in the usage patterns and the length of the viewing sessions. However, tablets are like mini-TVs and are seeing much longer viewing times than smartphones. In mobile TV, audio quality and audio synchronization with the video are particularly important since this helps viewers to follow the plot in situations when image quality is degraded.

1.3.3 Over-the-Air Digital TV

Consumer desire for media content anywhere and anytime continues to stress telecom wireless networks' ability to deliver video via their one-to-one architecture. As a result, cellular operators are imposing data caps on services. These data limits, together with the size and power limitations of personal devices, place significant constraints in the delivery of high-quality videos. This has raised concerns among content providers and consumers. On the other hand, the one-to-many architecture of over-the-air digital television (DTV) has an edge in delivering video to mass audiences with few constraints. The migration of analog to DTV has led to high-quality broadcast TV transmission. The service is available on big screen HDTVs as well as mobile devices. More importantly, the broadcast service provides a good source of live sports programs and allows emergency public information to be disseminated quickly. Many consumers are turning in to free DTV service for news and sports to supplement online movies.

All HDTVs come with mandatory in-built DTV receivers. In addition, mobile DTV is gaining momentum with the ratification of the Advanced Television Standards Committee (ATSC) mobile/handheld standard in September 2009. Portable network adapters that allow DTV signals to be decoded are commercially available for phones and tablets. Due to the limited over-the-air bandwidth, DTV may not be able to support UHD transmission. Although mobile DTV remains an "after-market" phenomenon, requiring dongles and other appendages to access the service on mobile devices, consumers may prefer this free service over cellular networks.

1.3.4 Non-Real-Time TV Delivery

The ATSC has developed a standard for the delivery of non-real-time (NRT) services over legacy and mobile DTV systems. Many TV programs do not need to be delivered in real-time but can be downloaded and presented when the viewer wishes to see them. NRT service is especially attractive for mobile TV viewing, which is normally unpredictable and done on an on-demand basis. Thus, the concept of "appointment viewing" may not always be practical. NRT service enables the consumer to pick what they want to see from a menu, with the program or service preloaded on their mobile device. NRT transmissions are file-based designed to be stored in the user's device until the user chooses to access them. A recent ATSC standard allows NRT video and audio content to be encoded using H.264/AVC and high-efficiency advanced audio coding (HE-AAC) respectively.

1.3.5 NRT Use Cases

Use cases for NRT include the following:

- Clipcasting – short-form video and audio clips similar to podcasts;
- VOD – may include short-form content, sports or news programs, music videos, standard length TV programs or full-length movies;
- Micro website – offers a similar experience to browsing the Internet using a Web browser where predefined content is downloaded to user's device;
- Out-of-home content and advertising – digital signage with multiple components such as live simulcast of the mobile TV channel, news and information content, advertising;
- Mobile emergency alert system (EAS) – disaster areas, evacuation routes, harsh weather, missing persons.

1.3.6 Cable Wi-Fi Alliance

In contrast to wired networks, wireless networks operate with lower bandwidth, variable latency, and occasional disconnections that cause coverage partitions. In particular, cellular networks will need to work in tandem with wire line backhaul networks in order to overcome the bandwidth crunch associated with video transmission. A cheaper and simpler alternative is to combine Wi-Fi with a broadband wire line connection. Unlike cellular networks that employ expensive and limited licensed radio spectrum, Wi-Fi networks rely on unlicensed spectrum with broader channels that can be aggregated. The latest Wi-Fi standard (ratified as 802.11ac in December 2013) is able to support a bit rate of up to 866 Mbit/s using a single transmitting antenna. This rate is significantly higher than 4G cellular and allows Wi-Fi to support compressed UHD video streaming.

The Cable Wi-Fi Alliance [2], a consortium of US cable operators (Bright House, Cox, Optimum, Time Warner Cable, Comcast), was formed to allow broadband Internet subscribers from any operator to access over 400,000 public and residential Wi-Fi hotspots. This number is further supplemented by additional hotspots from the cable operator. For example, Comcast subscribers can access millions of Wi-Fi hotspots in public areas and new neighborhoods that are powered by Comcast gateways. Currently, 9 million of such hotspots are available in the United States. Since most user data traffic is now generated indoors (due to data caps imposed by cellular operators), these Wi-Fi hotspots provide value-added, nomadic (rather than high mobility) wireless service for fixed broadband customers.

The new Wi-Fi gateways transmit two SSIDs using two independent antennas: a private one that is used by subscribers and a public one that enables neighborhood hotspots. Custom apps from the cable operators enable their subscribers to stream video and audio from the gateways to their mobile device. These apps demonstrate how watching TV becomes more convenient when it moves to the Internet, truly enabling TV anywhere. By combining a wired broadband Internet connection with high-speed Wi-Fi, the streaming performance may easily surpass 4G. Although the Wi-Fi network coverage and reach may be more limited than 4G, users are able to enjoy consistent high-speed access even when traveling outside the home. In addition, virtually any handheld device can be connected, including cheaper devices that do not have cellular connectivity. The concept is similar to BT-FON's Wi-Fi community networks, where free Wi-Fi gateways are given to subscribers to enable shared broadband Internet service.

1.4   SCREEN AND VIDEO RESOLUTION

It is useful to distinguish between screen and video resolutions, which can be different. The screen or display resolution relates to the total number of picture elements (pixels or pels) or dots (in computer lingo) that can be packed onto the physical size of the screen or display. We shall refer to these pixels as screen pixels. The video (or image) resolution relates to the total number of pixels in a single video frame. We shall refer to these pixels as video pixels. The actual video is often scaled (or resampled) to fit the physical size of the screen. If the size of the screen is much bigger than the video resolution, the video quality may be degraded if the video is scaled to the full screen since each video pixel will require either a greater number of identical screen pixels or larger screen pixels for display. Conversely, the video quality normally remains unaffected when a high-resolution video is mapped to a small screen. However, parts of the video may be cropped. Although TV screen resolutions have improved over the years, they tend to lag behind still-frame image resolutions. For instance, the resolution of a 3840×2160 UHD TV is roughly two times lower than a 16 megapixel digital camera that is available in some smartphones.

1.4.1 Aspect Ratios

The original video resolution may be varied by changing the pixel aspect ratio of the screen. The pixel aspect ratio of the screen is the ratio of the length to the width of the pixel on display. A pixel aspect ratio of 1:1 corresponds to square pixels whereas a ratio of 64:45 corresponds to rectangular pixels. The screen aspect ratio is related to its pixel aspect ratio by the governing relation: screen aspect ratio = original video resolution $\times$ pixel aspect ratio of screen. For example, if the original video resolution is 720/576, a screen aspect ratio of 16/9 can be obtained by multiplying 720/576 with a pixel aspect ratio of 64/45. Thus, each video image is displayed as a grid of rectangular pixels. This is normally the case if an image is mapped to a screen whose aspect ratio is different from the image. As a result, the image may look stretched or squashed in either the horizontal or vertical directions. For example, a circle may appear like an ellipse. However, the perceived video quality may not be affected.

1.4.2 Video Resolution

A list of common video resolutions is shown in Table 1.1. The lower resolutions (quarter common intermediate format (QCIF), common intermediate format (CIF), and standard definition (SD)) have been largely superseded by high-definition (HD) resolutions. The 720p, 1080i, and 1080p HD resolutions are by far the most popular whereas the UHD-1 resolution is emerging. A UHD-1 image offers four times the number pixels and twice the horizontal and vertical resolutions of a conventional 1080p image. Similarly, the UHD-2 resolution contains four times the number

Table 1.1   Common Video Resolutions.

Format	Resolution (pixels)
UHD-2 4320p (progressive)	7680×4320 (16:9 aspect ratio)
UHD-1 extra wide (progressive)	5120×2160 (21:9 aspect ratio)
UHD-1 2160p (progressive)	3840×2160 (16:9 aspect ratio)
High definition (HD) 1080p (progressive)	1920×1080 (16:9 aspect ratio)
High definition (HD) 1080i (interlaced)	1920×1080 (16:9 aspect ratio)
High definition (HD) 720p (progressive)	1280×720 (16:9 aspect ratio)
Standard definition (SD)	$720 \times 480, 720 \times 576$
Common intermediate format (CIF)	$352 \times 240, 352 \times 288$
Quarter common intermediate format (QCIF)	$176 \times 120, 176 \times 144$

of pixels of UHD-1. The majority of TV vendors and some service providers have primarily focused on the UHD-1 resolution. The difference in UHD-1 and UHD-2 video quality may not be significant. However, the difference in UHD-1 and 1080p video quality can be significant, especially at short viewing distances. Nevertheless, video publishers continue to target resolutions under 1080p for connected TVs, tablets, and phones to maximize compatibility and playability.

HD video may contain 720 or 1080 vertical progressive scan lines (720p or 1080p), 1080 interlaced scan lines (1080i), and is capable of displaying a 16:9 image (with square pixels) and output one or more channels of digital audio. Only progressive scan lines are allowed for UHD and new video coding standards such as H.265/HEVC. Although the aspect ratio is currently specified at 16:9 by ITU-R Recommendation BT.2020, many UHD TVs support an aspect ratio of 21:9. With progressive scan, all active lines are displayed in each video frame whereas in interlaced scan, odd and even lines are displayed in successive frames at half the frame rate. Clearly, the disadvantage of interlacing is that the horizontal resolution is reduced by half, and the video is often filtered to avoid flicker and motion rendering artifacts. Interlacing relies on the phosphor persistence of a cathode ray tube (CRT) screen to blend successive frames together. Since light emitting diode (LED) TVs and liquid crystal displays (LCDs) are rapidly replacing CRT screens, interlaced content must be deinterlaced at playback time. However, a screen with poor deinterlacing can result in a jittery image. Thus, deinterlacing is sometimes performed before video encoding. The interlaced video format remains popular with broadcast and pay-TV services (480i for SD, 1080i for HD) whereas the progressive format has been widely adopted by online video portals. The frame rate is normally specified together with the scan type. For example, a 1080p video with a frame rate of 60 Hz or 60 frames per second (fps) may be specified as 1080p60. The frame rates for UHD include 24, 50, 60, and 120 Hz. The new iPhone 6 is able to capture videos at 240 Hz.

Unlike common video formats, which are usually defined in terms of the vertical resolution (e.g., 720, 1080), digital cinema formats are usually defined in terms of horizontal resolution. These resolutions are often written as multiples of a base value of 1024 pixels using a "K" or "Kilo" suffix. Thus, a 2K image will have a length of 2048 pixels. A 4K format comprises 4096 × 2160 pixels whereas an 8K format comprises 8192 × 4320 pixels. UHD follows a similar format as digital cinema with a slight modification – a minimum horizontal resolution of 3840 pixels is specified. Since 3840 is exactly two times of 1920 and 2160 is exactly two times of 1080, a UHD-1 resolution of 3840 × 2160 allows convenient upscaling for 1080p HD formats. Thus, although the terms 4K and UHD are often used together, the first term originates from digital cinema whereas the second term is created for HDTVs. Both Samsung and LG Electronics have demonstrated 105-in. curved 5K UHD TVs with a resolution of 5120 × 2160 and an aspect ratio of 21:9. The new models are extra wide and almost six times sharper than 1080p HDTVs. The increase in spatial resolution may require a corresponding increase in the frame rate due to camera panning. These bendable TVs employ moving gears behind the TV panel to shift it from a flat screen position to one that curves the sides toward the viewers.

A UHD-2 image of 7680×4320 pixels contains 33 million pixels or 16 times as many pixels as a 1080p image. However, the resolution is still much lower than the processing power of 126 million pixels of a single human eye. An experimental UHD-2 digital video format was developed by NHK Science and Technology Research Laboratories [3]. The frame rate is 60 Hz progressive and the bandwidth is 600 MHz, giving a bit rate that ranges from 500 Mbit/s to 6.6 Gbit/s. Several years ago, NHK demonstrated a live satellite relay over IP for display over a 450-in. (11.4 m) screen. Video was compressed from 24 Gbit/s to a rate of about 100 Mbit/s whereas 22.2 channels of surround sound audio was compressed from 28 Mbit/s to roughly 7 Mbit/s.

1.4.3 Visual Quality

In addition to the screen and video resolution, other key factors that impact video quality include the number of pixels per inch (ppi) and the quantization level used in video compression. The ppi relates to the physical spatial resolution or pixel density of the screen. The ppi metric can be applied to any screen size and tends to be lower for devices with larger screens. For example, the maximum ppi for the iPhone and the iPad are 401 and 326, respectively. In general, a smartphone or tablet may be more suited for displaying 720p than 1080p or QCIF videos. This is because the finer details of a 1080p image may not be discernible on a small screen. On the other hand, a low resolution QCIF video may appear grainy. Thus, although many tablets and smartphones can support 1080p or higher, the majority of video publishers prefer to limit the resolution to 720p or lower for optimized playback across all devices rather than the best screen resolutions possible.

The number of bits representing each pixel (i.e., the color depth) and the quantization level for compressing the video also impacts video quality. For instance, 256 color levels are available for an 8-bit pixel. A typical video codec breaks the video image into discrete blocks (e.g., 8×8 pixels). These blocks are digitally transformed to the frequency domain, both horizontally and vertically. The values of the resulting block are then rounded off according to the quantization level. A coarse quantization level will lead to less accurate rounding and additional loss of color information. Thus, the video quality of a decompressed HD video may suffer if the quantization level is chosen for the compression is too coarse. On the other hand, a CIF video coded with a fine quantization level can achieve decent video quality.

1.4.4 Matching Video Content to Screen Size

Just as Web pages need to be redesigned for handheld devices, matching the nature of the video content to the screen size is important for the user experience. For example, watching a 3D video in the movie theater is a much better experience than watching the same movie on a smartphone. However, watching a 30-min cartoon or comedy may work well for the smartphone. Moreover, mobile users seldom watch full-length movies on 4-in. displays but will watch trailers, music videos, news conferences, and sports highlights. Thus, short videos are perfect for informing and entertaining mobile users. Clearly, visual effects, storyline, and video content all play a role in enhancing the user experience for multiscreens.

1.5   STEREOSCOPIC 3D TV

Several stereoscopic 3D (S3D) networks have emerged following the success of James Cameron's blockbuster science fiction 3D movie *Avatar* in 2009, surpassing another movie epic *Titanic*. However, S3D programming has not gained significant traction with viewers because of the higher price of 3D TVs and the requirement that viewers wear active glasses. These glasses act like shutters that display one of two 3D images for each eye to give the perception of depth. S3D sports was supposed to be a success but due to limited viewer adoption of S3D services to the home, ESPN shut down ESPN 3D, the first 3D network, in June 2013. The cost of S3D filming is incrementally higher because it requires additional cameras and crew. More importantly, a 3D TV has a more narrow viewing range than conventional TVs and requires the viewer to sit or stand upright. Wearing glasses can be uncomfortable or inconvenient for TV viewing. Family and friends may not be able to watch a game together – everyone will have to wear glasses and only a few people will be at an optimal angle to the screen to get a good viewing experience. These glasses are fairly expensive if they need to be replaced or if additional pairs are needed. Newer displays place a device called a parallax barrier in front of an image source, such as a LCD, to allow it to show a stereoscopic image without the need for the viewer to wear 3D glasses. A drawback is that the viewer must be positioned in a well-defined spot to experience the 3D effect.

1.5.1   Autostereoscopic 3D

Autostereoscopic 3D allows S3D images to be displayed without the use of glasses. This is normally achieved by employing two or more pairs of S3D images. Glass-free autostereoscopic 3D screens are used in a limited number of TVs, game consoles, and smartphones. For instance, the HTC Evo 3D smartphone comes with an autostereoscopic 3D touchscreen. The phone's 3D camcorder captures 720p videos (in addition to 2D 1080p videos), which are stored in full-resolution temporal format for easy preview and playback. Game consoles such as Nintendo 3DS also employ autostereoscopic 3D screens. Glass-free UHD TVs are also emerging and have been demonstrated.

1.5.2   Anaglyph 3D

Anaglyph 3D requires passive viewing glasses and generally offers wide-angle 3D viewing with less visual discomfort than S3D. Unlike S3D, which requires stereoscopic displays that are synchronized with active viewing glasses, a key advantage of anaglyph 3D is the compatibility with legacy TVs and regular displays. This means one can watch anaglyph 3D videos on a smartphone using cheap passive glasses.

These glasses allow vertical or horizontal polarized light to enter the left or right eye. Alternatively, clockwise or counter clockwise circularly polarized light may enter the left or right eye. Google's Street View employs anaglyph 3D videos, allowing users to view the streets in 3D.

1.6   VIDEO CODING STANDARDS

The Joint Video Team (JVT) is a partnership between the ITU-T (SG16) Video Coding Experts Group (VCEG) and the ISO/IEC (JTC 1/SC 29/WG 11) Moving Picture Experts Group (MPEG). JVT developed the H.264 or MPEG-4 Part 10 advanced video coding (AVC) video coding standard, which is widely deployed in many online video streaming systems. It succeeds the equally popular H.262/MPEG-2 standard. The emergence of UHD video services and the ubiquity of online video streaming have created a demand for coding efficiencies superior to H.264/AVC. To this end, MPEG and VCEG worked together to develop a new H.265 or High-Efficiency Video Coding (HEVC) standard. H.265/HEVC targets to improve the coding efficiency of H.264/AVC by at least two times under the same level of video quality. A comparison of the typical video compression efficiencies of MPEG coding standards is illustrated in Figure 1.2. H.265/HEVC supports technologies that enable parallel processing of large blocks of video data. Given its hierarchical block structure and significantly increased number of coding parameters, H.265/HEVC presents many new challenges, including coding optimization, mode decision, rate-control, and hardware design.

1.6.1   Exploiting Video Content Redundancies

The key objective behind video coding is to exploit redundancies that may be present within a video frame as well as in a sequence of video frames. If such redundancies

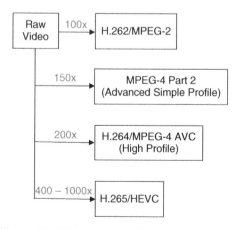

Figure 1.2   Efficiencies of video coding standards.

are present, only incremental changes are coded. For example, a pixel and its neighbors in a frame may be spatially correlated and such redundancy can be intracoded or intrapredicted. Similarly, pixels a video sequence may be temporally correlated in successive frames and such redundancy can be intercoded or inter-predicted. These redundancies are encoded as residual prediction errors and are employed by the decoder to predict current frames from previously decoded frames. Motion compensation provides additional coding gains for interprediction by using motion vectors to efficiently represent any motion of objects between successive frames. The differences between the original and predicted pixels are transformed and quantized. The final step in the coding process is entropy encoding, which exploits the redundancy in data representation to further compress the binary data using the fewest bits possible. The decoder follows the same steps in reverse but is computationally less complex than the encoder (typically 5–10 times simpler). In general, a HD video may require more time to encode (compress) and decode (decompress) than a SD video because more pixels need to be processed. Similarly, a UHD video will require more compression resources than HD videos.

1.6.2   High-Quality Versus High-Resolution Videos

HD interfaces such as HDMI send decompressed or decoded videos to the display. When a video is decompressed, the overall size of the video will be the same as the original raw video. This is because each pixel in each raw and decompressed video frame is represented by a fixed value, typically ranging from 0 to 255 using 8 bits. Hence, the frame sizes will remain the same. However, the video quality of the raw and decoded videos may differ since the decoded video may have undergone a lossy compression process.

It is important to distinguish between high-quality and HD videos. A high-quality video is encoded with a fine quantization level and it need not be a HD video. This is because the video quality of a HD video can be poorer than a SD video if the HD video is quantized at a very coarse level, which leads to significant information loss. In general, we refer to videos encoded with a fine quantization level as high-quality videos. These videos typically require a higher network bandwidth for transmission and a longer time to download because more color details from the original video are retained during the compression process, especially if the video resolution is high. Thus, transporting high-quality videos over bandwidth-constrained networks (e.g., cellular networks) can be a challenge.

1.6.3   Factors Affecting Coded Video Bit Rates

Coding efficiency, video content, frame rate, and quantization all impact the video bit rate. If coarse quantization is used, the quantization parameter (QP) value becomes larger. A larger QP value leads to a higher degree of lossy encoding, which reduces video quality and bandwidth requirements. Figure 1.3 shows the bit rates for a compressed video trailer. Video trailers are ideal for streaming because the initial sections of the video are usually simple scenes, allowing the receiving device to quickly buffer

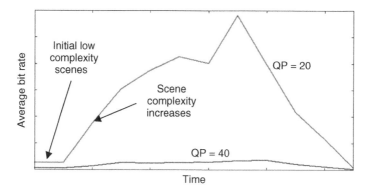

Figure 1.3 Bit rates for a compressed video trailer.

the frames for playback before more complex scenes arrive. The video bit rate variation is minimized for a QP of 40 and becomes content independent because more information is removed from each compressed video frame. As a result, the video quality may suffer, especially for more complex scenes. However, the low bit rates improve streaming performance and reduce the need for caching. Refining the quantization level using a lower QP of 20 improves the video quality but this is done at the expense of greater bit rate variation as well as significantly higher peak rates, which must be supported by the network and managed with sufficient caching at the receiver. These rates can be smoothed (or averaged) to a lower rate by buffering some video frames at the source before transmitting them, at the expense of greater delay jitter.

Many sports programs typically require higher frame rates to capture fast movements and this further increases the video bit rate. For example, a 60 Hz video requires twice the bit rate of the same video with a frame rate of 30 Hz. It is more challenging to transport UHD videos over a network. However, because consumer electronics have progressed rapidly with UHD TVs becoming available, both pay-TV and OTT providers are making UHD content available to their subscribers. To summarize, the variability of the compressed video bit rate is highly dependent on the quantization level and this should be appropriately chosen based the content type, frame rate, and the available network bandwidth.

1.6.4 Factors Affecting Coded Frame Sizes

The variation of the coded frame sizes is tightly correlated with the video resolution and content type. A higher resolution video tends to produce a broader range of coded frame sizes (Figure 1.4). For a fixed resolution, high-motion video content tends to produce a broader range of coded frame sizes (Figure 1.5). This implies that the amount of caching for the coded frames can be difficult to predict for high-resolution and high-motion videos. There is however, no correlation between the duration of the video and the frame size distribution. Thus, a longer video may not imply a higher frame size variability compared to a shorter video with the same resolution.

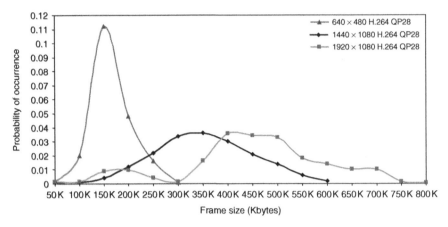

Figure 1.4   Coded frame sizes for videos with different resolutions.

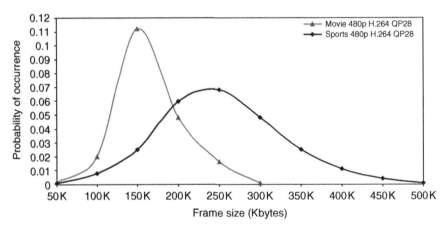

Figure 1.5   Coded frame sizes for videos with different content.

1.7   VIDEO STREAMING PROTOCOLS

Streaming video entertainment is the most efficient and convenient way compared to stored media using DVDs and old video home system (VHS) tapes. Streaming protocols determine how well compressed video can be transported across the network and the memory requirements at the user device. They simplify digital rights management (DRM) or content protection since small pieces of the video content are continuously delivered to the user device. In addition, the content is cached at the user device for a short period of time, thereby reducing the risk of content piracy for the entire video or movie. This makes it more acceptable to stream a video directly to a personal device (such as a PC, phone, tablet) and not via a proprietary STB, which employs hardware controls and encryption to prevent someone from reusing the video content. While streaming is the most popular mode of watching video, implementing trick modes (i.e., DVR functions such as fast forward and replay) can be challenging due to the network propagation delay. There are situations where video downloads may be more appropriate (e.g., downloading movies to watch on a plane). DVDs provide the best video quality, which is why they are still in demand in spite of the popularity of streaming.

1.7.1 Video Streaming over HTTP

Real-time Transport Protocol (RTP) over User Datagram Protocol (UDP) is a unidirectional IP-based network protocol designed for real-time multimedia traffic. Because UDP is a connectionless protocol that may result in packet losses and has difficulty passing through firewalls, it is normally employed by private managed networks (e.g., pay-TV cable networks) where packet losses are low. On the other hand, online video streaming is typically achieved using HTTP, a two-way connected-oriented network protocol that resends any data packet that is corrupted or dropped during transit. Thus, HTTP guarantees that all data will eventually be delivered correctly to the destination. Highly compressed video is sensitive to information loss, especially random packet losses caused by network congestion in the public Internet. Thus, loss-free protocols are more desirable and practical for video streaming over the Internet, although the end-to-end delay has to be calibrated properly for live video service. HTTP is a Web delivery standard that has become ubiquitous because it can work with standard Web servers and existing infrastructure, including CDNs, caches, firewalls, and NATs. The protocol can scale to a large number of media streams and is driven by clients that send GET requests to the servers and URLs that define the locations of the media data. Enhanced security for streaming can be achieved via secure protocols such as HTTPS.

1.7.2 Adaptive Bit Rate Streaming

HTTP's congestion control and reliability may overly burden real-time video streaming whereas UDP's nonreliable connections can affect user experience. Adaptive streaming is able to minimize the shortcomings of these protocols by dynamically adapting the video quality to the network bandwidth, connection quality, and device capabilities, thereby maintaining smooth playback. For example, when the client device detects network congestion or low bandwidth at any instant in time, it can request the server to deliver the video content at a reduced bit rate by selecting a stream with lower video quality. This prevents choppy playback when video frames do not arrive on time at the receiver (due to buffering at the sender) as well as serious video artifacts such as image breakup (due to packet losses). Thus, playback interruption is minimized when a client's network speed cannot support the quality of the video.

Clearly, the adaptive streaming process may increase the latency because the server and the player will need to switch to a lower quality video and make the necessary adjustments. Depending on the severity and frequency of the network congestion, this latency can sometimes exceed several seconds. However, because the coded video content is segmented into smaller chunks, it may be faster to switch chunks than change video coding parameters. Unlike peer-to-peer video streaming (e.g., Skype), which may consume large amounts of bandwidth in both inbound and outbound channels regardless of network conditions, adaptive streaming is more network friendly and generally consumes video bandwidth on the inbound. Adaptive streaming employs HTTP and is originally pioneered by Move Networks for online streaming of sports events. Apple, Microsoft, and Adobe have also developed their own approaches with new operating systems that support adaptive streaming and a compatible media player.

1.7.3 Benefits and Drawbacks of Adaptive Streaming

Adaptive streaming obviates users having to make choices between high- or low-quality video streaming and to adjust the video quality when switching from small to full screen mode. It also simplifies and reduces the cost of multiscreen (e.g., phone, tablet, PC, TV) deployment for service providers. The main drawback of this approach is a noticeable change in video quality as well as reinforced artifacts whenever there is a sudden degradation or improvement in network bandwidth or connection quality. If such instances occur frequently, it may compromise end-user quality of experience. On a private managed network, bandwidth availability and connection quality are more reliable. However, the use of adaptive streaming can still help maintain good video quality while smoothing the peaks that are associated with variable bit rate videos. For example, peak rates due to complex or high-motion scenes can be lowered by switching to a lower video quality for a short period of time. Thus, precise bandwidth dimensioning is not needed when adaptive streaming is applied to managed networks.

1.7.4 HTTP Progressive Download

Adaptive streaming relies on HTTP's rate control mechanism to determine the quickest sending rate and the appropriate level of video quality. To achieve fine-grained control of the video transmission, a progressive download mechanism based on HTTP can be employed. In this case, the user device may begin playback of the media before the download is complete. Low bit rate or short-form videos tend to be smaller in size and rather than streaming the videos continuously or changing the video quality levels, they can be downloaded instead when a broadband connection is available. Progressive download is widely used for the download and playback of short-form videos (e.g., YouTube) with a higher burst at the beginning to reduce the startup delay for playback. However, this method of video transmission may not allow the video to be fast forwarded quickly.

1.7.5 HTML5

HTML5 adds many new features such as audio, video, and canvas elements as well as support for scalable vector graphics (SVG). These features are designed to make it easy to add and handle multimedia and graphical Web content without relying on proprietary plugins and APIs. The canvas element enables dynamic, scriptable rendering of 2D shapes and bitmap images. Unlike canvas, which is raster-based, SVG is vector-based where each drawn shape is represented as an object and then rendered to a bitmap. This means that if attributes of an SVG object are changed, the browser can automatically re-render the scene. A canvas object would need to be redrawn.

1.8 TV INTERFACES AND NAVIGATION

Accessing digital movies should be as simple as flipping a channel. The STB has been the bedrock of cable and IPTV operators for over two decades. The box scans for available TV channels and allows the user to select the desired channels via a TV remote. A tablet may work better as a user and social TV interface than a remote. Several years ago, Comcast developed an app to allow tablets to act as a TV remote and communicate with the STBs. The app connects the tablet to the STB using the enhanced TV binary exchange format (EBIF), the cable industry's specification for delivering one-to-one interactive applications to STBs. This pioneering personalized remote allows users to view channels, browse the full channel lineup, and invite friends to watch movies. With the emergence of Internet TV, STBs have been gradually replaced by video streaming boxes and HDMI adapters that enable connections to HDTVs. Using a broadband connection, media entertainment can be streamed via Wi-Fi using these devices. There are a variety of streaming boxes (e.g., Apple TV, Roku, Amazon Fire TV) and adapters (e.g., Chromecast, Roku). These devices also allow smartphones and tablets to stream content to the TVs.

1.8.1 Streaming Adapters

With a thumb-sized Chromecast HDMI dongle plugged into a TV, online entertainment (e.g., movies, TV shows, and music from Netflix, YouTube, HBO GO, Hulu, Google Play Movies and Music, Pandora, Vudu, Crackle, Rdio) and any content from the Google Chrome Web browser can be streamed wirelessly from any Wi-Fi device to the TV. Users select the media to play using Chromecast-enabled mobile apps and Web apps, or via a beta feature called "tab casting" that can mirror content from the Chrome browser tab. The apps also allow users to control playback and volume. Android and Apple mobile apps are supported using an extension. Chromecast employs a system-on-a-chip, which includes video codecs for hardware decoding of VP8 and H.264/AVC. Like Chromecast, which conserves space, the Roku HDMI adapter provides access to Netflix, Hulu, Amazon Instant Video, ESPN, HBO Go, Showtime, YouTube, Redbox Instant, among more than 1000 music, movie, TV, and sports channels.

1.8.2 Streaming Boxes

The Apple TV streaming box provides Netflix, Hulu, ESPN, YouTube, Vevo, Vimeo, MLB.TV, NBA League Pass, NHL GameCenter, along with any music or videos purchased through Apple's iTunes. AirPlay allows users to broadcast anything on their iOS device to the HDTV through the box. Users can also play games with dual screens, leveraging the TV and either an iPhone, iPad or iPod Touch. The Roku streaming box provides access to more than 1000 music, movie, TV and sports channels, including Netflix, Hulu, Amazon Instant Video, ESPN, HBO Go, Showtime, YouTube, and Redbox Instant. The remote includes a headphone jack for private listening, and features games including Angry Birds and classics such as Pac-Man or Galaga. Amazon has launched a powerful Fire TV streaming box with a voice-activated remote, providing instant access to over 200,000 TV episodes and movies, plus all subscriptions and streaming services, including Amazon Instant Video, Netflix, Hulu, ESPN, Vevo, Bloomberg, and Showtime.

1.8.3 Media-Activated TV Navigation

The last digital island, the TV, has joined the PC, tablet, and smartphone as an Internet-connected device. The broadband TV brings all of the Internet to the TV, including the full range of video websites. This development has a profound impact on the distribution and consumption of digital media. Media-activation can help improve the navigation, selection, and playback of videos, photos, and music. As an example, Google Talk personalizes the Web search engine with speech input from the user. Google also provides a voice-recognition application that converts speech commands to actions via devices with speech input such as a smartphone. Media-activated communications can also be nonverbal. For example, popular immersive games employ video sensors on game consoles to recognize human body movement while hand gestures can also be analyzed to navigate TV or video thumbnails.

1.8.4 Smartphone and Tablet TV Navigation

Smartphones and tablets can serve as TV remotes (more sophisticated but better), allowing the user to browse for video content, control playback, and adjust volume. Such a capability is more flexible than the Internet-connected TV, where users can only watch online content from preprogrammed video portals. In addition, navigation using a TV remote is less intuitive than using a phone or tablet with a familiar keyboard. Thus, the widespread deployment of Internet TV may eventually remove the traditional remote as well as Internet/cable connections for the TV, which may potentially evolve to become a display-only device.

1.8.5 Digital Living Network Alliance

The Digital Living Network Alliance (DLNA) [4] has developed open technical specifications to ensure interoperability among TVs, STBs, Blu-ray players, and other mobile devices (e.g., tablets, phones) so that these devices can seamlessly stream multimedia content to each other over point-to-point wireless (e.g., 802.11) or wired (e.g., Ethernet, MoCA) connections. DLNA employs the Universal Plug and Play

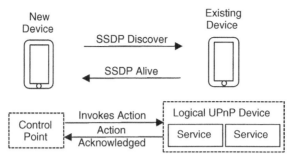

Figure 1.6   SSDP discover message flow.

(UPnP) networking protocols that enables networked devices to seamlessly discover each other on the network and establish functional network services for media sharing and communications. UPnP is the network analogy of plug-and-play.

The Simple Service Discovery Protocol (SSDP) forms the basis of the discovery protocol of UPnP. SSDP is a network protocol that allows HTTP clients and services to discover each other without depending on server-based configuration mechanisms, such as the Dynamic Host Configuration Protocol (DHCP) or the Domain Name System (DNS), or relying on static configuration of a network device. SSDP defines the communication protocols between control points and devices that provide services. A UPnP device may contain both functions. A typical message flow for discovering services between UPnP devices is shown in Figure 1.6.

1.8.6   Discovery and Launch

Discovery and Launch (DIAL) is a mechanism for discovering and launching applications on a local area network such as a home network. It enables second screen devices (e.g., phones, tablets) to send content to first screen devices (e.g., TVs, Blu-ray players, streaming boxes and adapters). DIAL is a protocol co-developed by Netflix and YouTube with help from Sony and Samsung. It is used by Chromecast and relies on UPnP.

1.8.7   UltraViolet

Unlike DVDs, which play on any DVD device, consumers are not comfortable with limited usage of online video content. For example, videos that are compatible with Flash may not be compatible with Windows Media or Apple QuickTime. Leading entertainment and consumer-electronics companies have therefore formed a consortium, the UltraViolet to develop technical specifications that content distributors and manufacturers can follow to ensure that consumers are not locked to a specific platform. The goal is to let consumers know that content and devices carrying a special logo will play seamlessly with one another.

REFERENCES

1. Netflix Open Connect, https://openconnect.netflix.com.
2. Cable Wi-Fi Alliance, http://www.cablewifi.com.
3. NHK, http://www.nhk.or.jp/strl/english.
4. Digital Living Network Alliance, http://www.dlna.org.
5. UltraViolet, https://www.myuv.com.

HOMEWORK PROBLEMS

1.1. Explain whether user interactivity, ease of use, content sharing, and video quality are important differentiators between online and pay-TV. Will the age of the consumer play a role in your assessment?

1.2. Explain whether Wi-Fi or DLNA enabled HDTVs are better than the use of HDMI streaming adapters such as Google's Chromecast. State-of-art STBs are becoming thumb-sized. Will they compete directly with streaming adapters?

1.3. Will autostereoscopic 3D require the viewer to sit or stand upright? Will it provide wider viewing angles than stereoscopic 3D?

1.4. Consider the following streaming method that sends video frames at a constant frame rate. Instead of sending a video at a bit rate that is based on the file size, we now send the video at the same frame rate that the video is meant to be played at. Typical frame rates are 30 and 60 Hz. A few frames are buffered by the receiver before they are played back. Will this method show good performance for constant frame rate videos? Consider another situation when videos are encoded with variable frame rates where fast moving scenes (such as action scenes) require more frames to be generated by the encoder over the same time interval. How will such variable frame rate videos impact the performance of the constant frame rate streaming method as outlined earlier?

1.5. Streaming a video at a constant frame rate may achieve good performance for constant bit rate videos. For such videos, the average bit rate for each frame remains more or less fixed and the natural frame rate of the video is held constant. This is achieved by adjusting the quantization level for every encoded frame to produce a constant output bit rate. With variable bit rate videos, both the frame rate and the quantization level (hence video quality) are held sconstant for all frames and so fast changing or complex scenes will require more bits per frame. Explain whether video streaming at a constant frame rate will achieve good performance for variable bit rate videos. Note that constant bit rate videos may not imply a constant sending rate at the source since there may be instances of network congestion or poor quality links that mandate the source to buffer or retransmit some of the video frames.

1.6. An alternative to adaptive streaming is to maintain an acceptable video quality suitable for the subscribed network bandwidth by encoding the video as efficiently as possible to ensure smooth playback. How will network congestion (that lead to packet dropping) impact adaptive and fixed video quality streaming? Can constant bit rate video encoding replace adaptive streaming or can they work together?

1.7. Will the new iPad be able to display UHD-1 videos? What is the best resolution for video display on a smartphone?

1.8. Suppose the same movie is recorded in analog VHS and digital DVD formats. Which format will generally produce better video quality? Does quantization play a role in the video quality? How do they compare to the video quality of pay-TV and online TV distribution? Note that in analog television, there are roughly 480 active lines with each line holding about 440 pixels. Thus, each video frame has slightly more than 200,000 color pixels. This should be compared with the resolution of the SD video, which has 720×480 or 345,600 pixels/frame.

1.9. In some US cities, high-speed fiber-optic Internet and TV access to the home is readily available. Will this development remove the need for video compression?

1.10. If adaptive streaming is implemented over managed pay-TV networks, will the user still be able to record all or part of the video on a DVR?

1.11. Will adaptive streaming lead to lower buffering and faster start times for the client device when compared to traditional constant video quality streaming?

1.12. When adaptive streaming starts, the client may request chunks from the lowest bit rate stream. If the client discovers that the download speed is greater than the bit rate of the downloaded chunk, it will request higher bit rate chunks. If the download speed deteriorates later, it will request a lower bit rate chunk. Suppose the highest bit rate stream is selected at the start instead. Discuss the tradeoffs for doing this.

1.13. What are the benefits of using fixed size versus variable size chunks in adaptive streaming? Note that ultimately, these chunks are transported by network technologies such as Ethernet and IP that employ variable size packets.

1.14. Discuss the roles of the following methods to improve video transport over a network: (a) buffering at the source, (b) buffering at the receiver, (c) segmentation of compressed video into chunks, and (d) encoding the video at constant bit rate.

1.15. Which of the following achieves the best video quality when displaying a 1080p video: (a) smartphone, (b) tablet, (c) HDTV, (d) UHD TV, and (e) giant TV screen at a football stadium? Note that the viewing distance is a key factor that affects perceived video quality.

1.16. Why do consumers still prefer Netflix even though the service does not offer live sports, unlike Web portals from service providers?

1.17. Will the traditional remote enable more effective discovery of video content than a personal device such as a phone or tablet?

1.18. Analyze why Internet protocols (e.g., Real-time Transport Streaming Protocol (RTSP), Real-time Transport Protocol, (RTP), Real-time Transport Control Protocol, (RTCP)) that are designed with real-time guarantees for transporting multimedia traffic may not perform better than HTTP. RTP facilitates packet resequencing and loss detection using sequence numbering of packets, synchronization between multiple flows using the RTP timestamp and RTCP Sender Reports, and identification of the payload type. Evaluate the advantages of using RTP for video streaming over the public Internet.

1.19. Suppose a user employs a tablet to video chat with a friend who is using a smartphone. Is adaptive streaming useful for such an application? Suppose the same user now connects the tablet to an OTT provider to watch a movie. Explain whether the streaming requirements for this application will be different to the video chat application. Note that the first video application is an example of human–human communications (a live application, possibly interactive) whereas the second video application is an example of machine–human communications (an on-demand application). Are there video applications that involve machine–machine communications?

1.20. Periodic traffic is broadly defined as a traffic type where the average bit rate can be determined. This is in contrast to bursty traffic where the average rate is unpredictable and may therefore cause packet losses due to network congestion. Such congestion is often solved by adding buffers, at the expense of increased end-to-end latency. Delay jitter is a key metric that determines the useful playback lifetime of periodic traffic whereas fairness in network bandwidth consumption is a key metric for bursty traffic. Explain whether adaptive video traffic should be classified as periodic or bursty traffic.

1.21. Many consumers will watch the best available content on the best available screen at the best available video quality and price. Rank these four criteria in the order of importance and justify your answer.

1.22. Cloud gaming requires graphics steaming, where game objects are represented by 3D models and textures, and are streamed to the players' devices, which perform rendering of the scenes. Alternatively, video streaming can be employed, where the cloud not only executes the game logic but also the game rendering, and streams the resulting game scene to the players' devices as video. Evaluate the strengths and weaknesses of each of these methods by balancing bandwidth and delay limitations with wider accessibility and possibility to run the game on thin clients.

1.23. One reason why music piracy is more prevalent than video piracy is because compressed audio requires much smaller storage space and bandwidth requirements. Will next-generation video codecs such as H.265/HEVC with vastly improved coding efficiency create DRM problems in video distribution?

1.24. Suppose a compressed video quantized with QP 28 is decoded to raw video and then compressed again using QP 28. Will the original and resulting file sizes and video quality match up?

1.25. Compute the pixel aspect ratio to convert an old 4:3 video to 16:9. Compute the screen dimensions of a 401 ppi smartphone to display a 1080p video using square pixels.

2

VIDEO CODING FUNDAMENTALS

Compression reduces the cost of storing and transmitting a video by converting the information to a lower bit rate. This chapter describes the main building blocks of video coding and many practical tips on improving the compression efficiency and video quality for different types of content. These fundamental principles underpin legacy as well as emerging video coding standards that support efficient video delivery and storage.

2.1 SAMPLING FORMATS OF RAW VIDEOS

Storage space and transmission bandwidth dictate the need for video compression. Unlike audio, raw or uncoded videos require massive storage space. For example, a raw 60-min SD video may require over 100 Gbytes of storage, which is roughly equivalent to 25 standard DVDs. A raw 1-min UHD-1 video may take up as much as 50 Gbytes. A raw UHD-2 video doubles the storage requirements. Each digital video frame, which may also be known as a picture or an image, is sampled and represented by a rectangular matrix or array of picture elements called pixels or pels for short. The term pixel is commonly used to describe still-frame images. If the image contains smooth surfaces (e.g., sky, grass, walls), the colors of adjacent pixels can be highly correlated (i.e., pixel correlation approaches 1).

Each raw pixel in a video frame can be separated into three samples that correspond to three different color components. They are the luminance (Y) sample and

Next-Generation Video Coding and Streaming, First Edition. Benny Bing.
© 2015 John Wiley & Sons, Inc. Published 2015 by John Wiley & Sons, Inc.

two color samples namely red chrominance (C_r) and blue chrominance (C_b). The YC_rC_b components are also collectively known as the YUV sampling format or color space. The luminance (luma) sample represents brightness whereas the chrominance (chroma) samples represent the extent in which the color deviates from gray toward red or blue. Each sample of the color component is represented by fixed integer value, typically ranging from 0 to 255 for 8 bits of precision (0 for white, 255 for black). Thus, the luma and chroma components of a video frame can be represented by three rectangular matrices (planes) of integers. By defining a color space, samples can be identified numerically by their coordinates. The term sample (rather than pixel) is more commonly used in video coding standards since the luma and chroma samples may require a different set of coding parameters.

2.1.1 Color Subsampling

The luma sample usually takes precedence over the chroma samples because the human eye is more sensitive to brightness than color (e.g., hue, saturation). Thus, the sampling resolution of both chroma samples can be reduced. For instance, if both chroma samples are subsampled by a factor of 2 in the horizontal and vertical directions, this gives rise to the 4:2:0 color format. In this case, each chroma component has one-fourth of the number of samples of the luma component (half the number of samples in horizontal and vertical dimensions). The 4:2:0 format is by far the most popular and has been adopted by DVD and Blu-ray players as well as new video coding standards such as H.265/HEVC. If a 4:2:0 video frame is progressively sampled with a rectangular size of $W \times H$ (where W is the width and H is the height of the frame in terms of luma samples), then the rectangular dimension of each chroma component array is reduced to $W/2 \times H/2$. Thus, a 4:2:0 16×16 luma block will contain two 8×8 chroma blocks.

Table 2.1 shows the various sampling formats and the raw efficiencies. Although 8-bit samples are used in Table 2.1, the bit depth for the samples can be extended to 10, 12 or even 14 bits. An increased bit depth improves the compression efficiency for each coded sample due to the greater precision. In 4:2:0 sampling, the C_r and C_b matrices are half the size of the Y matrix in both horizontal and vertical dimensions. In 4:2:2 sampling, the C_r and C_b matrices are half the size of the Y matrix in the horizontal dimension but the same size in the vertical dimension. In 4:4:4 sampling,

Table 2.1 Color YC_rC_b (YUV) Formats and Raw Efficiencies.

YUV 4:4:4	Typically 8 bits per Y, U, V plane	24 bits/pixel
	No subsampling	8 bits/sample
YUV 4:2:2	$4Y$ samples for every $2U$ and $2V$	16 bits/pixel
	2:1 horizontal subsampling	8 bits/luma sample
	No vertical subsampling	
YUV 4:2:0	$4Y$ samples for every $2U$	12 bits/pixel
	2:1 horizontal subsampling	8 bits/luma sample
	2:1 vertical subsampling	

the size of the C_r and C_b matrices is the same as the Y matrix in both vertical and horizontal dimensions. A 4:0:0 sampling format contains only the luma component and no chroma components. The format is used in monochrome (black and white) videos.

2.1.2 YUV Versus RGB Color Space

The YUV color space is often used in broadcast TV. The red, green, blue (RGB) trichromatic color space is more popular for computer graphics. Unlike RGB, brightness (Y) information is encoded separately from color information in YUV. In general, an RGB video achieves better compression efficiency than YUV due to the greater correlation between RGB components. The RGB space can be converted to YUV via standard mathematical formulas. For example, the Y component can be derived from the RGB components as shown in equation (2.1), where $0.2125 + 0.7154 + 0.0721 = 1$. The individual RGB values may change depending on the video quality:

$$Y = 0.2125R + 0.7154G + 0.0721B \qquad (2.1)$$

This relation is valid for modern LED or LCD HDTVs, smartphones, and tablets. It places more emphasis on the green component because the human eye is more sensitive to that wavelength of light in the perceived brightness. The U and V components can be derived from equations (2.2) and (2.3):

$$U = B - Y \qquad (2.2)$$

$$V = R - Y \qquad (2.3)$$

2.1.3 Bit Rate and Storage Requirements

For most videos, including high definition (HD) videos, the color depth is 8 bits. For ultra high definition (UHD), the color depth can be 10 or 12 bits. Table 2.2 shows the

Table 2.2   Bit rates and Storage Requirements for 4:4:4 8-bit Color Format Videos at 30 Hz That Are Encoded with an Efficiency of 0.25 bit/pixel.

Video Resolution (Progressive)	Raw Bit Rate (Mbit/s)	Encoded Bit Rate (Mbit/s)	Storage for 1-h Encoded Video (Gbyte)
352×288 (CIF)	73.0	0.76	0.34
720×480 (SD)	248.8	2.59	1.17
1280×720 (HD)	663.6	6.91	3.11
1920×1080 (HD)	1493.0	15.55	7.00
3840×2160 (UHD-1)	2986.0	31.10	14.00
7680×4320 (UHD-2)	5972.0	62.20	28.00

encoded bit rates and storage requirements. In general, HD over H.262/MPEG-2 is very expensive to deliver, typically requiring four times the bit rate over SD. Thus, the increasing popularity of HD videos drives the adoption of more efficient standards such as H.264/AVC and H.265/HEVC. For example, an UHD-1 video can be compressed to 15 Mbit/s using H.265/HEVC, which is roughly equivalent to the rate of a H.262/MPEG-2 HD video.

2.2   IMPACT OF VIDEO COMPRESSION

The primary aim of video compression is to remove spatial and temporal redundancy so as to encode a video at the minimum bit rate for a given level of video quality or improve the video quality for a given bit rate. The redundancy is inherent in the video content because on the average, a small number of samples change from one frame to the next. Hence, if only the changes are encoded, a significant amount of storage space or network bandwidth can be conserved. The video quality of a compressed video is largely dictated by the encoding process. Lossy video coding improves the compression efficiency (i.e., smaller compressed file size) compared to lossless compression (e.g., data compression).

Many video coding methods are lossy where information that is removed cannot be recovered. This leads to irreversible distortion of the original video. The distortion can be minimized because information that the humans cannot perceive is typically removed. On the other hand, lossless video coding will not impact video quality when the video is decoded for display. In many instances, lossless coding is not used since the coding efficiency is lower compared to lossy coding. In addition, videos that are coded losslessly may take a longer time to decode due to the larger frame sizes. The best video quality is obtained when the source video is captured as a raw video without lossy compression. Many video capture devices, including professional video production equipment, will compress the source video to some degree. Because the lossy compression process is noninvertible, degradation in video quality may be inevitable, although this loss is usually minimized by the capture device. Video service providers may increase the loss further with more aggressive compression in order to conserve bandwidth for network transport.

2.2.1   Rate-Distortion Optimization

The coded video bit rate can be reduced by allowing some acceptable distortion in the video. This process is known as rate-distortion optimization (RDO). RDO allows a vendor to differentiate its encoder products. RDO algorithms essentially choose the encoding modes to achieve the best trade-off between low distortion and low bit rate. Figure 2.1 shows a typical trade-off between the coded video bit rate and the distortion. The coded rate can be limited to a maximum rate at the expense of some distortion. Alternatively, a maximum level of distortion can be defined and the coded rate can be minimized. The peak coded rate is obtained for lossless coding but there is no distortion. These constrained optimization problems can be complex. The

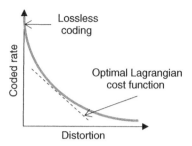

Figure 2.1   Rate-distortion optimization.

Lagrangian optimization method removes these individual constraints (i.e., maximum coded rate or distortion level) and uses a cost function C as defined in equation (2.4) for joint optimization of R and D. The Lagrange multiplier (λ) is usually dependent on the quantization parameter (QP). The distortion metrics are typically based on the sum of squared differences (SSD) and the sum of absolute differences (SAD).

$$C = D + \lambda R \tag{2.4}$$

where D = distortion level, R = coded video bit rate, λ = Lagrange multiplier (independent of D and R).

When the cost function is rewritten as equation (2.5), it corresponds to a series of parallel lines with a constant slope of $-1/\lambda$. This requires a convex rate-distortion curve for optimization. The optimal Lagrange cost function is the line (with negative slope) that is tangent to the convex hull as shown in Figure 2.1. The resulting optimized value of λ (say $\lambda_{\min}$) is the same for all coded blocks. Large values of λ correspond to high quality encoding.

$$R = -\frac{1}{\lambda}D + \frac{1}{\lambda}C \tag{2.5}$$

2.2.2   Partitions in a Video Frame

A video frame is normally divided into smaller blocks of samples before compression is applied. The size of these blocks may range from superblocks (64×64 samples) and macroblocks (MBs) (16×16 samples) to even smaller partitions (8×8, 4×4 samples). Each block is coded either spatially or temporally. The MB is a fundamental building block of legacy video coding standards (e.g., H.262/MPEG-2, H.264/AVC) and contains a section of the luma component and the spatially corresponding color or chroma components. A 4:2:0 MB consists of 6 blocks ($4Y, 1C_b, 1C_r$), a 4:2:2 MB consists of 8 blocks ($4Y, 2C_b, 2C_r$), and a 4:4:4 MB consists of 12 blocks ($4Y, 4C_b, 4C_r$). Note that the baseline block is 8×8 samples. Hence, $4Y$ is equivalent to 16×16 luma samples and $1C_b$ and $1C_r$ are 8×8 samples each. Thus, each 4:2:0 MB comprises 16×16 luma samples and two 8×8 chroma samples whereas for 4:2:2 and 4:4:4 MBs, the dimensions for the chroma samples are 8×16 and 16×16 respectively.

2.2.3   Video Coding Standards

Many video coding standards specify only the bitstream syntax and the decoding process. Thus, they do not guarantee any coding efficiency. New video coding standards emphasize a high level of parallelism, which refers to the ability to simultaneously process multiple regions of a single frame. Support for such parallelism is useful to both encoders and decoders when multiple computing cores can be used concurrently. However, because video compression is designed to remove redundancy associated with content dependency, this makes parallelism challenging.

2.2.4   Profiles and Levels

Profiles and levels define subsets of the video bitstream syntax and semantics that indicate the decoder capabilities to decode a certain bitstream. A profile specifies a set of coding tools that are enabled or disabled in generating conforming bitstreams. Profiles are needed to ensure that the encoded video plays properly on the target decoder with the minimum set of functions. For example, a video coded using the H.264/AVC baseline profile will decode properly on any baseline profile decoder and additional H.264/AVC main profile functions are not needed. If identical coding parameters are chosen, the profile has no impact on the video quality, coding time, or coded video file size.

Unlike software decoders, hardware decoders are more restrictive on the profile they support. This is not surprising since the number of implemented features impacts decoder complexity and cost. Minimizing the number of profiles enhances interoperability between devices. While profiles define coding capabilities, levels restrict other properties of the encoded video, including parameters of the bitstream such as resolution, bit rate, frame rate (for a given resolution), and number of reference frames. These constraints impose restrictions on values of the syntax elements in the bitstream. The number of reference frames and frame size must be consistent with the level.

2.3   GENERAL VIDEO CODEC OPERATIONS

The general operations of a video codec are shown in Figure 2.2. The encoding process is typically more time consuming than the decoding process. While many videos can be pre-encoded in advance, the decoding latency should be reasonable for streaming applications and channel switching in broadcast operations. Each video frame is typically partitioned into a grid of blocks, which are analyzed by the encoder to determine the blocks that must be transmitted (i.e., blocks that contain significant changes from frame to frame). To do this, each block is spatially or temporally predicted. Spatial or intraframe prediction employs a block identifier for a group of samples that contain the same characteristics (e.g., color, intensity) for each frame. The identifier is sent to the decoder. Temporal prediction predicts interframe motion and compensates for any inaccuracy in the prediction. The difference in the frame information, which also corresponds to the prediction error, is called a residual

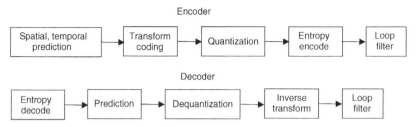

Figure 2.2   General encoder and decoder functions.

2.3.1   Transform Coding

The residual is transformed on a block-by-block basis into a more compact form. The residual and prediction information, such as the prediction mode and motion vectors (MVs), are sent to the decoder to reconstruct the original video frame. To improve the quality of the intracoding, the edges of the previously encoded blocks of the current frame may be applied to the intracoded block. These previously encoded blocks may or may not be intracoded. The primary task of the transformation process is to reduce the spatial redundancy or correlation that may be present in the residuals. Subsequent coding steps such as quantization and entropy coding are then designed to compress such coefficients.

The transform comprises orthogonal basis functions. When the transform is applied block-wise, it produces as many coefficients as there are samples in the block (i.e., there is no compression yet). Although transform coefficients take up the most bandwidth, they can be compressed more easily because the information is statistically concentrated in just a few coefficients. This process is called transform coding, which reduces the inherent spatial redundancy between adjacent samples. Subsequently, the transform coefficients of a block are quantized using a scalar quantizer.

2.3.2   Quantization

Quantization achieves further compression by representing the coefficients with no greater precision than is necessary to achieve the desired video quality. The orthogonal transform bases enable distortion metrics such as SSD to be computed directly in the transform domain when determining the appropriate quantization level for the transform coefficients. The spacing between a range of original coefficient values that corresponds to a quantized or reconstructed value can be of equal length (i.e., uniform), which is useful for fine-grained quantization. Uniform quantization is commonly employed in video coding standards, including H.262/MPEG-2, H.264/AVC, and the emerging H.265/HEVC standards. For more efficient quantization, however, nonuniform spacing can be used, as shown in Figure 2.3. The quantized transform coefficients may be rounded, scanned (usually in a zigzag or field fashion), scaled (binary-shifted), and then entropy encoded. Like the quantization steps, an unequal or nonuniform number of bits can be assigned to the quantized values to enable more

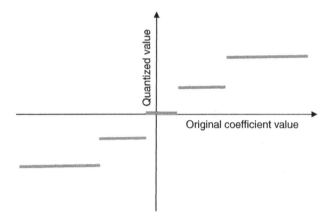

Figure 2.3   Nonuniform scalar quantization.

efficient entropy encoding using variable length codes (VLCs). For example, shorter codes can be assigned to more likely quantized values and the probability of each value is determined based on the values used for coding surrounding blocks. In addition to the transform coefficients, entropy encoding is used to losslessly compress other syntax elements such as motion vectors and prediction modes. It is the final step in video encoding and the first step in video decoding. At the decoder, the inverse of the video encoding process is performed except for the rounding.

The information loss in quantization is essentially determined by the QP, which affects the accuracy of the quantization matrices. The loss due to quantization can be managed using RDO. For example, the appropriate QP can be chosen based on the overall distortion level of the video frame and the block activity level. The QP and the Lagrange multiplier are usually held constant for all coded blocks of the video frame when determining the amount of loss. For example, the Lagrange multiplier can be calculated using equation (2.6). The amount of loss that can be tolerated depends on the video resolution and the distortion perception threshold of the human visual

QP = 20   QP = 40

Figure 2.4   A lower QP leads to better video quality.

system, which can vary from subject to subject. In general, a smaller QP reduces the loss, which leads to a corresponding improvement in video quality (Figure 2.4). However, this increases the encoding time and degrades the video coding efficiency. The size of the video buffer determines the variability of the video quality of the encoded video. A high and uniform video quality requires a large buffer, which implies a large encoding delay:

$$\lambda = 0.85 \times 2^{(QP-12)/3} \tag{2.6}$$

2.3.3 Deblocking Filter

In-loop deblocking filtering is applied after inverse quantization and transform before the reconstructed frame is used for predicting other frames using motion compensation. The filtering is done as part of the prediction loop rather than postprocessing. In this way, the frames will serve as better references for motion-compensated prediction since they incur less encoding distortion. Since block edges are typically reconstructed with less accuracy than interior samples, the filter operates on the horizontal and vertical block edges within the motion compensation loop in order to remove artifacts caused by block prediction errors. For example, if there is a large absolute difference in sample values near a block edge, it may lead to a blocking artifact. The filtering reduces the bit rate while producing the same objective quality as the unfiltered video. In addition, the subjective quality of the decoded video is improved because the blockiness is reduced without compromising the sharpness.

Deblocking filters are highly adaptive. The strength of the filter can be adapted to the difference in the sample values near the block edge and is controlled by several syntax elements. The threshold that controls the strength of the filter is quantizer dependent because blockiness becomes more pronounced with coarse quantization. This implies that with a good deblocking filter, a higher QP can be employed, which leads to more bit rate savings. It is important to distinguish real edges in the frame and those created by quantization of transform coefficients. Real edges should be left unfiltered as much as possible. If the absolute difference between samples near a block edge is relatively large, a blocking artifact is quite likely. However, if the absolute difference is so large and cannot be linked to the coarseness of quantization, a real edge is more likely and should not be filtered.

To separate the two cases, the sample values across every edge are analyzed. Suppose $\{s_3, s_2, s_1, s_0\}$ and $\{t_0, t_1, t_2, t_3\}$ are the samples of two adjacent 4×4 blocks. The actual boundary is between s_0 and t_0. Filtering of s_0 and t_0 takes place if their absolute difference falls below some threshold α and the absolute sample differences on each side of the edge (i.e., $|s_1 - s_0|$ and $|t_1 - t_0|$) fall below another threshold β where $\beta \ll \alpha$. To enable filtering of s_1 or t_1, the absolute difference of $|s_2 - s_0|$ or $|t_2 - t_0|$ has to be smaller than β in addition to the first two criteria. The thresholds α and β are dependent on the quantizer. Thus, the filter strength is linked to the general quality of the reconstructed frame prior to filtering. For small quantization values, both thresholds become zero and filtering is disabled.

2.4   TRANSFORM CODING

Transform coding is a key component of video compression. A good transform is able to decorrelate the input video samples and concentrate most of the video information (or energy) using a small number of transform coefficients. In this way, many coefficients can be discarded, thus leading to compression gains. The transform should be invertible and computationally efficient, such as the ability to support subframe video coding using variable block sizes. In addition, the basis functions of a good transform should produce smoother and perceptually pleasant reconstructed samples. Many video coding standards employ a block transform of the residuals. Intracoded blocks and residuals are commonly processed in $N \times N$ blocks (N is usually 4, 8, 16, 32) by a two-dimensional (2D) discrete transform.

The use of smaller transform sizes such as 4×4 leads to several benefits as intra and intercoding methods improve. This is because the residual contains less spatial correlation, which implies the transform is less effective in statistical decorrelation. Smaller transforms also generates less noise or ringing artifacts around the block edges and require fewer computations. The 2D transformation is typically achieved by applying one-dimensional (1D) transforms in the horizontal (i.e., row-wise) and vertical (i.e., column-wise) directions. The elements of the transform matrix can be derived by approximating scaled integer basis functions of the discrete transform under considerations such as limiting the dynamic range for transform computation and maximizing the precision and closeness to orthogonality when the matrix elements are specified as integer values.

2.4.1   Orthonormal Transforms

The transform of an $N \times N$ block of video samples (s_{original}) can be represented in a separable form by applying an $N \times N$ orthonormal transform (A) using equation (2.7). In this case, A^T is the transpose of the transform, which is identical to the inverse of the transform (A^{-1}) if the transform is orthonormal. Since the orthonormal transform contains symmetric matrix elements, this requires fewer mathematical operations than an ordinary matrix multiplication. The inverse transform is shown in equation (2.8):

$$s_{\text{transform}} = As_{\text{original}}A^{-1} = As_{\text{original}}A^T \qquad (2.7)$$

$$s_{\text{original}} = A^T s_{\text{transform}}A \qquad (2.8)$$

Since adjacent blocks tend to have similar properties, one may predict the block of interest from the surrounding blocks, typically the ones located on top and to the left of the block of interest, since those blocks have already been encoded. The difference between the actual block and its prediction is then coded, which results in fewer bits to represent the block of interest compared with applying the transform directly to the block itself. Figure 2.5 shows an example of how a block of 8×8 samples may employ the boundary samples of a previously encoded block for intraprediction. Samples 1, 9, 17, 25, 33, 41, 49, and 57 can be set to the same value as sample A if they are predicted

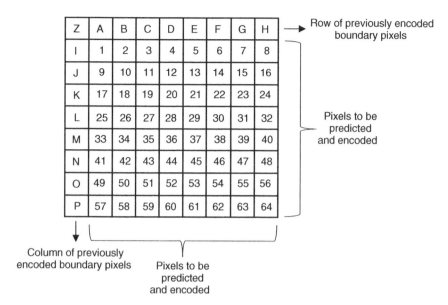

Figure 2.5 Intraprediction for a block of 8×8 samples.

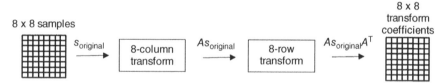

Figure 2.6 Transform coding for a block of 8×8 samples.

vertically downward. Other intraprediction directions (e.g., horizontal, diagonal) can also be employed.

The block of 8×8 image samples can be converted to the frequency domain using an orthonormal transform. The process produces 64 transform coefficients (which are generally decorrelated due to the transform process) as shown in Figure 2.6. Each individual block is typically coded using two modes. An intracoded block is transformed and the coefficients are quantized and entropy coded. For an intercoded block, a motion vector (MV) that specifies the translation from a corresponding block in a previously coded reference frame is first obtained. This MV and the transform coefficients of the residual block are quantized and coded.

The quantized transform coefficients correspond to different frequencies, with the coefficient at the top left-hand corner of the block representing the DC value. High-frequency components, such as those produced by differential coding of inter-coded frames, are typically quantized coarsely. On the other hand, intracoded contain information in a broad range of frequencies and, therefore, require fine-grained quantization. The fidelity of chroma coefficients is improved by using finer quantization step sizes compared to the luma coefficients, especially when the luma coefficients are quantized coarsely. It is well known that images of natural scenes have predominantly low-frequency components because the values of spatial and temporally adjacent samples vary smoothly, except in regions with sharp edges. In addition, the human eye can tolerate more distortion to the high-frequency components than to the low-frequency components.

2.4.2 Discrete Cosine Transform

The discrete cosine transform (DCT) expresses a finite sequence of discrete data points as a sum of cosine functions oscillating at different frequencies. DCT is widely applied to lossy compression of video and images, and is well suited for image areas of roughly uniform patterns and complexity. DCT is similar to the discrete Fourier transform (DFT) but uses only real numbers. Hence, all DCT coefficients are real coefficients. The use of cosine rather than sine functions is critical for compression because cosine functions are more efficient. DFT typically produces poorer energy compaction and more blocky artifacts when compared to DCT. Instead of sending all samples in a frame, DCT packs the energy of the samples into a few frequency coefficients, which are transmitted. The residuals may differ depending on the quality of the prediction and whether the prediction exploits spatial or temporal redundancy. If there is no spatial prediction from samples of adjacent blocks, the energy compaction ability of DCT may approach the Karhunen–Loève transform (KLT), which produces completely decorrelated transform coefficients and achieves optimal energy compaction for the video samples. Unlike the KLT, which has no simple computational algorithm and is seldom used in practice, DCT matrices are usually quite small because they are able to handle single sharp-edged blocks of samples. A deblocking filter is required to smooth the edges between blocks.

DCT can be applied to each row and column of the block. The transform produces a coefficient matrix where the top-left $(0,0)$ element is the DC or zero-frequency coefficient. Entries with increasing vertical and horizontal index values represent higher vertical and horizontal spatial frequencies. DCT exhibits a favorable energy compaction property on the intra and residual video data. Thus, most of the information tends to be concentrated in a few low-frequency components of the DCT. In lossy video coding, small high-frequency components can be discarded. This is usually the case when coding video images, which contain much correlation within each frame and between frames. Thus, only a few frequency coefficients are transmitted instead of the entire block of sample values. The 1D DCT and inverse DCT formulas are shown in (2.9) and (2.10), where x_n denotes a sample value. In video coding, a 2D DCT is normally applied to a square block of pixels. The 2D DCT transform and inverse transform for a $N \times N$ block are shown in (2.11) and (2.12), where $x_{n,m}$ is a sample value assigned to the block in the (n, m) location:

$$X_k = \sum_{n=0}^{N-1} x_n \cos\left[\frac{\pi}{N}\left(n + \frac{1}{2}\right)k\right] \qquad k = 0, 1, 2, \ldots, N-1 \qquad (2.9)$$

Table 2.3   Sample Values of a Residual Block.

				N				
	158	158	158	163	161	161	162	162
	157	157	157	162	163	161	162	162
	157	157	157	160	161	161	161	161
m	155	155	155	162	162	161	160	159
	159	159	159	160	160	162	161	159
	156	156	156	158	163	160	155	150
	156	156	156	159	156	153	151	144
	155	155	155	155	153	149	144	139

$$x_n = \frac{1}{2}X_0 + \sum_{n=1}^{N-1} X_k \cos\left[\frac{\pi}{N}\left(k + \frac{1}{2}\right)n\right] \qquad k = 0, 1, 2, \ldots, N-1 \qquad (2.10)$$

$$X_{j,k} = \sum_{n=0}^{N-1}\sum_{m=0}^{N-1} x_{n,m} \cos\left[\frac{(2n+1)j\pi}{2N}\right]\cos\left[\frac{(2m+1)k\pi}{2N}\right] \qquad j, k = 0, 1, 2, \ldots, N-1$$

$$(2.11)$$

$$x_{n,m} = \frac{1}{N^2}\sum_{n=0}^{N-1}\sum_{m=0}^{N-1} X_{j,k} \cos\left[\frac{(2n+1)j\pi}{2N}\right]\cos\left[\frac{(2m+1)k\pi}{2N}\right] \qquad j, k = 0, 1, 2, \ldots, N-1$$

$$(2.12)$$

Table 2.3 shows the sample values of an 8 × 8 residual block. The values are highly correlated because they fairly similar with only minor differences. Transform coding generally removes this correlation. However, for motion-compensated predicted coded blocks, the residuals may not always be as correlated and this may compromise the energy compaction efficiency of DCT. The DCT coefficients are scanned as an array in ascending order, starting from low to high-frequency coefficients in a zigzag diagonal fashion, as shown in Figure 2.7. The statistical distribution of the transform coefficients typically show large values for the low-frequency part and decreasing to small values for the high-frequency part. For this reason, coefficient values are coded in reverse scan order (i.e., starting from small to large coefficient values). The positions of nonzero transform coefficients are transmitted together with their values.

After the transformation, Table 2.4 is obtained. As can be seen, there are very few high-valued DCT coefficients relative to the coefficient centered in the (0, 0) or DC location. The values in Table 2.4 can be normalized (so that most of the values are less than 1) and quantized, giving rise to Table 2.5. Since most of the coefficients are zero, they are not coded or transmitted. Run-level coding of the format (zero-run, value) can be employed, where zero-run is the number of coefficients quantized to zero starting from the last nonzero coefficient and value is the amplitude of the current nonzero coefficient. In this case, only six nonzero coefficients are coded. Thus, the amount of data that can be conserved is factor of 64/6 or about 10 compared to the case when all 64 coefficients are sent. The size of the residual can be an individual block (as in this

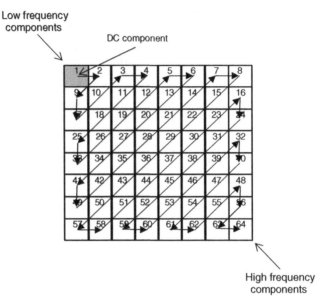

Figure 2.7 Zigzag DCT coefficient scanning.

Table 2.4   Transformed DCT Coefficients.

				J				
	1259.6	1.0	−12.1	5.2	2.1	1.7	−2.7	−1.3
	22.6	−17.5	6.2	−3.2	2.9	−0.1	−0.4	−1.2
	−10.9	9.3	−1.6	−1.5	0.2	0.9	−0.6	0.1
K	7.1	−1.9	−0.2	1.5	−0.9	−0.1	0.0	0.3
	−0.6	0.8	1.5	−1.6	−0.1	0.7	0.6	−1.3
	−1.8	−0.2	−1.6	−0.3	0.8	1.5	−1.0	−1.0
	−1.3	0.4	−0.3	1.5	−0.5	−1.7	1.1	0.8
	2.6	1.6	3.8	−1.8	−1.9	1.2	0.6	−0.4

example) or an entire frame. The denormalized and inverse transformed coefficients are shown in Tables 2.6 and 2.7, respectively. The recovered residual is identical to the original.

There are several enhancements of DCT, which can be suboptimal for some types of residuals. To improve the coding of intraprediction residuals, directional cosine transforms can be employed to capture the texture of the block content. An example is the mode-dependent directional transform (MDDT), which is an approximation of KLT. In MDDT, the transform basis functions are dependent on the intraprediction direction, which can be derived by extensive training using the spatial and temporal activities of the video content. For instance, different vertical and horizontal transforms can be applied to each of the nine modes (of block size 4×4 and 8×8) in H.264/AVC intraprediction. Because the MDDTs are individually designed for each

Table 2.5   Normalized and Quantized DCT Coefficients.

			J					
	21	0	−1	0	0	0	0	0
	2	−1	0	0	0	0	0	0
	−1	1	0	0	0	0	0	0
k	0	0	0	0	0	0	0	0
	0	0	0	0	0	0	0	0
	0	0	0	0	0	0	0	0
	0	0	0	0	0	0	0	0
	0	0	0	0	0	0	0	0

Table 2.6   Denormalized DCT Coefficients.

			J					
	1260	0	−12	0	0	0	0	0
	23	−18	0	0	0	0	0	0
	−11	10	0	0	0	0	0	0
k	0	0	0	0	0	0	0	0
	0	0	0	0	0	0	0	0
	0	0	0	0	0	0	0	0
	0	0	0	0	0	0	0	0
	0	0	0	0	0	0	0	0

Table 2.7   Inverse Transformed Coefficients of Reconstructed Residual Block.

				N				
	158	158	158	163	161	161	162	162
	157	157	157	162	163	161	162	162
	157	157	157	160	161	161	161	161
m	155	155	155	162	162	161	160	159
	159	159	159	160	160	162	161	159
	156	156	156	158	163	160	155	150
	156	156	156	159	156	153	151	144
	155	155	155	155	153	149	144	139

prediction mode based on training data, they require the storage of 18 (9 horizontal and 9 vertical) transform matrices.

Transform coding can be skipped to improve the intracoding of certain types of video content, such as animated content, which tend to produce more high-frequency components with DCT. This will in turn, reduce the coding efficiency of DCT. The residuals are quantized directly in the spatial domain without applying any transform. Transform skip can also be extended to intercoded blocks.

2.4.3 Discrete Sine Transform

Discrete sine transform (DST) expresses a finite sequence of discrete data points as a sum of sinusoids with different frequencies and amplitudes. DST is an odd extension of the original function and is similar to the DFT of purely real functions. Unlike DFT, which uses both cosines and sines (in the form of complex exponentials), DST uses only sine functions. This leads to different boundary conditions for DST and DFT. Because DSTs operate on finite, discrete sequences, the function must be specified as even or odd at both the left and right boundaries of the domain and around what point the function is even or odd. Consider a sequence (a, b, c) of three equally spaced data points and an odd left boundary is specified. There are two possibilities for the odd extension: $(-c, -b, -a, 0, a, b, c)$ or $(-c, -b, -a, a, b, c)$. These choices lead to the variations of DSTs and also DCTs. Each boundary can be either even or odd (two choices per boundary) and can be symmetric about a data point or the point halfway between two data points (two choices per boundary), for a total of $2 \times 2 \times 2 \times 2$ or 16 possibilities. Half of these possibilities where the left boundary is odd correspond to the eight types of DST; the other half are the eight types of DCT.

2.4.4 Asymmetric DST

Spatial prediction of a block is normally achieved by employing previously encoded neighboring samples at block boundaries. After spatial prediction is performed, the energy of the residual samples close to one boundary of a residual block may differ significantly from the other end. Unfortunately, DCT does not take into account this energy variation across samples in the residual block. In fact, the basis functions of DCT maximize the energy at both edges of the block. To overcome this limitation and improve spatial intracoding, a DST variant can be used with basis functions that diminish at the known block boundary, while retaining high energy at the unknown boundary. Due to such asymmetric structure of the basis functions, the transform is called asymmetric discrete sine transform (ADST) [1]. A key requirement of ADST is the boundary information of previously encoded blocks must be known and reliable (undistorted).

Since the basis functions of the DCT achieve their maximum energy at block edges, this may lead to blockiness where incoherence in the quantization noise of adjacent blocks becomes magnified. This problem can be addressed by postfiltering using a deblocking filter at the decoder to smooth the block boundaries. However, the process can result in information loss when sharp details become blurred. Because the basis functions of ADST vanish at block edges with known boundaries, this removes the need for deblocking.

2.4.5 Comparison of KLT, ADST, and DCT

The relative coding gains of the ADST and DCT transforms with respect to KLT when applied to prediction residuals for different intersample correlation are shown

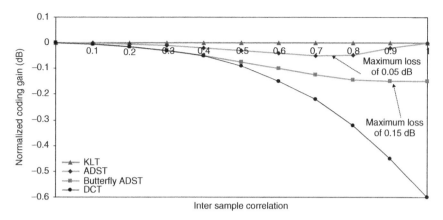

Figure 2.8   Relative coding gains versus intersample correlation for 8 × 8 blocks.

in Figure 2.8. A first-order Gauss–Markov model for the image signal with available partial boundary information is employed. For low or high intersample correlation, the distortion level of ADST may approach the minimum that is achievable with KLT. The distortion level of DCT also approaches that of KLT and ADST for low intersample correlation (such as intersample correlation of 0 or black-on-white samples). However, the distortion increases for high intersample correlation. Thus, ADST is preferred over DCT when the intersample correlation is high. The theoretical advantage of ADST over DCT is a 0.68 dB or 14% coding gain for a 4 × 4 block with an intersample correlation of 1 (i.e., best case using identical color samples). For a larger block dimension of 8 × 8, this gain drops to about 0.6 dB. This is due to the improvement in coding efficiency when DCT operates on larger blocks. A key drawback with the original ADST is that it does not allow butterfly implementation for fast parallel computations. Although butterfly versions of ADST have been proposed recently to enable parallel computations [2], the efficiency is compromised slightly (0.15 dB maximum) compared to the original ADST due to precision loss (integer rounding-off errors) in the approximation of the basis functions. In addition, the distortion level for butterfly ADST no longer approaches that of KLT at high intersample correlation.

The maximum coding gain for butterfly ADST versus DCT is about 0.43 dB or 10% for high intersample correlation and 8 × 8 blocks. This gain drops to between 0.1 and 0.2 dB (2–5%) for intersample correlation ranging from 0.5 to 0.6. If the intersample correlation falls below 0.4, there is no coding gain. In practice, the ADST coding gains are moderate. Although medium to high level of intersample correlation may exist in a video image, especially for high-resolution videos that contain fine details or videos that employ a greater color bit depth, this is offset by the need to encode the video using a low QP value to improve the reliability of the boundary information used by ADST. This, in turn, reduces the coding gains since the compressed videos are now larger. Hence, DCT may still be able to maintain a comparable level of coding efficiency as ADST.

2.4.6 Hybrid Transforms

Hybrid DCT/ADST transform coding leverages on the intrinsic benefits of DCT and ADST. The hybrid transform may switch between DCT and ADST, depending on the reliability and availability of block boundary information. For example, if the boundary information is available and reliable, ADST can be used. On the other hand, if the boundary information is unavailable or highly distorted, DCT can be used. This implies that DCT is a mandatory transform because blocks located at the edge of a video frame do not have access to boundary information. There are some implementation considerations. The use of a hybrid transform requires higher computational complexity to operate the two transforms as well as additional processing to determine the appropriate instance to switch transforms. This may be a challenge for video decoding in low-cost handheld devices with limited computing power. Moreover, the gain in coding efficiency for ADST/DCT may not be significant. Thus, although hybrid DCT/ADST has been adopted by WebM/VP9 for intracoding, other new video coding standards such as H.265/HEVC still prefer a single transform to handle the coding of residuals.

2.4.7 Wavelet Transform

Wavelet transform can be used in lossless or lossy video compression. It can be used to represent high-frequency components in 2D images and is particularly effective for compressing transient signals than DCT. Each DCT coefficient represents a constant pattern applied to the whole block whereas each wavelet coefficient represents a localized pattern applied to one section of the block. Thus, wavelet transforms are usually very large with the aim of taking advantage of large-scale redundancy in an image. Wavelet transforms employ variable length basis functions that typically overlap, thus avoiding the need for a deblocking filter. Wavelet transform tends to introduce blurry artifacts whereas block-based DCT tends to be characterized by blocky artifacts. However, wavelet-coded videos may suffer from poorer coding efficiency when the predicted data is in the form of blocks. In addition, because of the overlapped transforms, intraprediction is not possible. Thus, DCT has been shown to perform better than the wavelet transform at various image resolutions.

2.4.8 Impact of Transform Size

The DCT treats a block as periodic and has to reconstruct the resulting difference in sample values at the boundaries. For 64×64 blocks, a huge difference is likely at the boundaries, which require more high-frequency components for reconstruction. Larger transform sizes may enjoy several performance advantages over smaller transform sizes. For example, the time cost per sample of a 32×32 inverse transform is less than twice that of an 8×8 inverse transform. In addition, larger transforms may often exploit the fact that most high-frequency coefficients are typically zero. Larger transforms can be constructed by using smaller transforms as building blocks. Consider an 8×8 block consisting of four 4×4 blocks. If each 4×4 block is transformed separately, four DC components with the same precision are required. With an 8×8 block, only one DC component is needed but because the value is potentially higher, it may be subjected to higher quantization noise.

2.4.9 Impact of Parallel Coding

In traditional video coding, blocks are sequentially coded. Thus, when coding a particular block, available boundary information is limited to only a few (and not all) of its edges. However, new video coding standards are designed for parallel encoding and decoding, which is important for delivering HD/UHD services, including live events. In this case, smaller blocks of video data may be organized into larger tiles, which are independently encoded and decoded. This may degrade the compression efficiency because boundary information of previously encoded blocks may not be available to code a block. For example, some of the blocks lying at the boundary of the tile may not have access to boundary information. A compromise to this problem is to employ a reasonable number of coding blocks within the tiles to maximize the coding efficiency of the integer transform while leveraging on a high level of parallelism to reduce coding delay for the tiles. Note that parallel decoding may not be useful for handheld devices with limited processing power. For non real-time video content, turning off parallel processing may help improve the coding efficiency and minimize storage and transmission bandwidth requirements. Thus, the trade-offs for enabling parallel threads in encoding and decoding should be calibrated against the coding efficiency and the device/content characteristics.

2.5 ENTROPY CODING

Entropy coding is a lossless or reversible process that achieves additional compression by coding the syntax elements (e.g., transform coefficients, prediction modes, MVs) into the final output file. As such, it does not modify the quantization level. VLCs such as Huffman, Golomb, or arithmetic codes are statistical codes that have been widely used.

2.5.1 Variable Length Codes

VLCs may assign shorter code word to frequent bit strings and longer code word to less frequent bit strings, thus leading to unbalanced code trees with unequal probabilities. This is in contrast to fixed-length codes that lead to optimal balanced trees when the bit strings are equally probable. Contiguous bit strings can be represented as a single value and a multiplier. For example, a bitstream of 0000000000000000111111110000000 00000000000000000 can be represented as 0×16, 1×8, 0×24 or 0 [10000], 1 [01000], 0 [11000]. Thus, the original 48 bits can be transmitted using only 18 bits. This technique is particularly effective for encoding videos because there are often large fields of identical colors. The lossless compression ratio is typically less than 4:1. Entropy decoding becomes a bottleneck at high bit rates, especially for intracoded frames.

Table 2.8   Examples of Exponential Golomb Codes.

Code Number	Code Word
0	1
1	010
2	011
3	00100
4	00101
5	00110
6	00111
7	0001000
8	0001001
...	...

2.5.2   Golomb Codes

Golomb codes are special Huffman VLCs with a regular construction scheme that favors small numbers by assigning them shorter code words. These code words have the generic form of $\{K$ zeroes, 1, K-bit data$\}$ where data is a binary representation. Such codes are parameterized by a nonnegative integer K. The encoding procedure is a 2-step process. The data to be coded is first converted to binary except for the last K digits and 1 is added to it. The number of bits is counted and 1 is subtracted from it. That number corresponds to the number of 0 bits that must be appended to the start of the previous bit string, as shown in Table 2.8. This single code word table can be customized according to the data statistics and can be extended infinitely. Each code word can be decoded as Code Number $= 2^K + \text{integer(data)} - 1$. In entropy coding, each code word can be mapped to a syntax element using a single code word table. A different VLC table for each syntax element is avoided.

2.5.3   Arithmetic Coding Overview

Arithmetic coding encodes a group of binary symbols (bins) using a unique VLC so that the average bit rate is minimized. Unlike Huffman and Golomb coding, arithmetic coding does not require a code table. To enable efficient coding and representation of the bins, statistical dependencies are removed. When a bit string is arithmetic encoded, frequently used characters or bins are represented with fewer bits and not-so-frequently occurring bins are represented with more bits. Arithmetic coding differs from Huffman coding in that rather than separating the input into bins and replacing each bin with a code, arithmetic coding encodes the entire message as a fraction. Thus, bit strings are converted to single floating point numbers between 0 and 1. Any real number in the interval $[0, 1)$ can be represented in binary as $0.b_1 b_2 \ldots$ where b_i is a bit. Arithmetic coding begins by predicting the patterns to be found in the bins of the message. An accurate prediction optimizes the output. Adaptive models continually change their prediction of the data based on the pattern in the bitstream. The same model must be used by both encoder and decoder.

2.5.4 Nonadaptive Arithmetic Coding

In the simplest case, the probability of each symbol occurring is equal. For example, consider a set of three bins, A, B, and C, each equally likely to occur. Simple block encoding requires 2 bits/symbol, which is wasteful since one of the bit variations is never used (if A = 00, B = 01, C = 10, then 11 is unused). A more efficient solution is to represent a sequence of these bins as a rational number in base 3 where each digit represents a symbol. For example, the sequence ABBCAB becomes 0.011201_3. The next step is to encode this ternary number using a fixed-point binary number of sufficient precision to recover it, such as $0.0|01|01|10|0|01_2$. This requires only 10 bits. Thus, 2 bits are saved compared with block coding. To decode the value, the length of the original string must be known. The fraction can be converted back to base 3 and rounded to 6 digits to recover the string.

2.5.5 Steps in Nonadaptive Arithmetic Coding

The first step in arithmetic coding is to represent each string x of length n by a unique subinterval $[L, R)$ within the range $[0, 1)$. The width $r - l$ of the subinterval $[L, R)$ represents the probability of x occurring. The width of the subinterval is approximately equal to the probability of the string. The subinterval $[L, R)$ can itself be represented by any number, called a tag, within the half open subinterval. The k significant bits of the tag $0.t_1 t_2 t_3 \ldots$ represent the code of x. For example, if the binary tag is $0.t_1 t_2 t_3 \ldots = (L + R)/2$, then k is chosen to form the code $t_1 t_2 \ldots t_k$ and should be the shortest possible so that $L \leq 0.t_1 t_2 t_3 \ldots t_k 000 \ldots < R$. Figure 2.9 illustrates

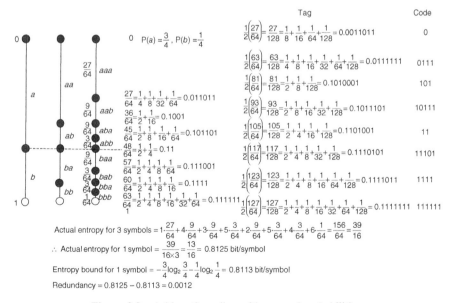

Figure 2.9 Arithmetic coding with unequal probabilities.

Table 2.9 Adaptive Coding for *aabac*.

	Initial	a	a	b	a	c
a	1	2	3	3	4	4
b	1	1	1	2	2	2
c	1	1	1	1	1	2
d	1	1	1	1	1	1

the procedures when unequal probabilities are assigned to the two subintervals corresponding to bins *a* and *b*. The entropy bound is a lower bound for the average length of the code word per symbol. As can be seen, the actual entropy of arithmetic coding is very close to the entropy bound, thus achieving virtually no redundancy. However, this assumes that the probabilities assigned to the bins are accurate and reflect the actual probabilities of occurrence in the bitstream.

2.5.6 Context-Based Adaptive Arithmetic Coding

New video coding standards tend to favor arithmetic coding over Huffman or Golomb coding because it is more flexible. For example, arithmetic coding can be made adaptive using a simple mechanism and requires only a small table of frequencies for every context. The context may relate to the structure of the transform coefficients. In contrast, Huffman coding requires a more elaborate adaptive algorithm as well as a coding tree for every context. Consider a one-symbol context. The probabilities for each context are maintained. For the first symbol, the equal probability model is employed. For each successive symbol, the model for the previous symbol is employed. To make this model adaptive, first initialize all symbols that have a frequency of 1. After a symbol is coded, its frequency is incremented by 1. The new model is used for coding the next symbol. As an example, suppose the character string *aabac* is encoded using adaptive arithmetic coding. After five iterations, the probability model is 4/9 for *a*, 2/9 for *b*, 2/9 for *c*, and 1/8 for *d* as shown in Table 2.9.

2.5.7 Code Synchronization

Because VLC code words are not of fixed-length, bit errors may cause desynchronization when the code words become corrupted. The bits following the error become undecodable but can be skipped until the next unique code word is recognized. A special resynchronization marker (e.g., 0x0001) or a slice header can be used to facilitate the process. The latter method is used in some video coding standards. By using a self-synchronization code word at the beginning of each row of blocks (known as a slice), the damage caused by any transmission error is restricted to a single row, so that the upper and lower blocks of a damaged block may still be correctly received. However, in some cases, even after synchronization is obtained, the decoded information may still be useless since there is no way to determine whether the information corresponds to spatial or temporal locations.

2.6 MPEG (H.26x) STANDARDS

The ISO/IEC Moving Picture Experts Group (MPEG) family of video coding standards is based on the same general principles: spatial intracoding using block transformation and motion-compensated temporal intercoding. The hierarchical structure supports interoperability between different services and allows decoders to operate with different capabilities (e.g., devices with different display resolutions). Popular software-based MPEG codecs such as FFMPEG [3] are readily available. Powerful H.264/AVC and H.265/HEVC encoders such as x264 and x265 are based on FFMPEG.

2.6.1 MPEG Frames

MPEG exploits the spatial and temporal redundancy inherent in video images and sequences. The video sequence is the highest syntactic structure of the coded bitstream. The temporal sequence of MPEG frames consists of three types, namely intracoded (I) frames, interpredicted (P) frames, and bidirectionally interpredicted (B) frames. The P and B frames are collectively known as intercoded frames. These frame types aim to strike a balance between random access of frames, compression efficiency, and maintaining good video quality. Some of these frames can also be used as references for interprediction.

2.6.2 I Frames

I frames are key frames that provide checkpoints for resynchronization or re-entry to support trick modes (e.g., pause, fast forward, rewind) and error recovery. They are commonly inserted at a rate that is a function of the video frame rate. These frames are spatially encoded within themselves (i.e., independently coded or intracoded) and are reconstructed without any reference to other frames. Since subsequent frames after the I frame will not reference frames preceding the I frame, I frames are able to restrict the impact of error propagation caused by corrupted frames, thus providing some form of error resiliency. Each frame is subdivided into blocks of samples and each block is encoded with respect to itself and sent directly to the block-based transform.

Intracoded I frames tend to suffer from reduced coding efficiency when compared to intercoded frames. However, because intracoding involves intraprediction of the different directions of correlation within the I frame, the coding efficiency is generally much better than using lossless data compression, which focuses only on redundancy within a bit string. The improvement in intracoding efficiency is due largely to individual video images containing some level of correlation within individual subjects (e.g., face, background, smooth surface). If these subjects are largely static, intercoded frames will provide additional gains. Of course, if the video image becomes more noise-like (common occurrence in black and white videos), intracoding efficiency will degrade.

2.6.3 P Frames

P frames are temporally encoded using motion estimation and compensation techniques. P frames are first partitioned into blocks before motion-compensated prediction is applied. The prediction is based on a reference to an I frame that is most recently encoded and decoded before the P frame (i.e., the I frame is a past frame that becomes a forward reference frame). Thus, P frames are forward predicted or extrapolated and the prediction is unidirectional. As an example, the residual can be calculated between a block in the current frame and the previously coded I frame. An MV is calculated to determine the value and direction of the prediction for each block.

2.6.4 B Frames

B frames are temporally encoded using bidirectional motion-compensated predictions from a forward reference frame and a backward reference frame. The I and P frames usually serve as references for the B frames (i.e., they are referenced frames). The interpolation of two reference frames typically leads to more accurate interprediction (i.e., smaller residuals) than P frames. The residual is calculated between a block in the current frame and the past/future reference frames. Due to more references, B frames incur higher bit overheads than P frames. For example, twice the number of MVs is needed. One MV is used to determine the magnitude and direction of the forward prediction whereas the other MV is used to determine the magnitude and direction of the backward prediction. However, because B frames are derived from past and future reference frames, they may achieve better compression efficiency, especially for low-motion or low-complexity video content.

B frames are typically omitted for constant bit rate (CBR) encoding (i.e., only I and P frames are employed) since their inclusion creates a difficult challenge when adapting the QP of each coded frame to match a targeted bit rate. B frames take a longer time to encode compared to I and P frames. Hence, the use of B frames should be minimized for live video content. Newer video coding standards allow B frames to be used as reference frames. For example, B frames are not used as reference frames in H.262/MPEG-2 but this is permitted in newer standards such as H.264/AVC and H.265/HEVC. The goal to eventually remove the distinction between P and B frames. However, B frames typically do not serve as a reference for other frames (i.e., they are unreferenced frames). Thus, they can be dropped without significant impact on the video quality. Such action may be necessary when a B frame becomes corrupted during transmission.

2.6.5 Intracoded P and B Frames

If the residual is large and motion compensation becomes ineffective, the encoder may intracode a block in a P or B frame. In other words, P and B frames are not always intercoded. Thus, some P or B frames may have large frame sizes.

2.7 GROUP OF PICTURES

MPEG video sequences are made up of groups of pictures (GOPs), each comprising a preset number of coded frames, including one I frame and one or more P and B frames. Pictures are equivalent to video frames or images. The I frame provides the initial reference to start the encoding process. The interleaving of I, P, and B frames in a video sequence is content dependent. For example, video conferencing applications may employ more B frames since there is little motion in the video. On the other hand, sports content with rapid or frequent motion may require more I frames in order to maintain good video quality. This implies that there may be little difference in the compression efficiencies of new and legacy video coding standards for sports content. Coupled with high frame rates, sports content typically require higher bit rates than any other content.

2.7.1 GOP Length

Longer GOPs tend to suit low-motion video content better because there is lower dependency on the I frames. This in turn improves the video compression efficiency. However, long GOPs may degrade error resilience, which may be a problem for streaming media and Blu-ray authoring. Longer GOPs also increase the latency in the transmission of the frames since the entire GOP must be assembled before transmission can occur. Low-motion content tends to allow more B frames in a GOP without sacrificing video quality. On the other hand, choosing long GOPs or GOPs with more B frames for high-motion or high-action movies with frequent scene changes may degrade the video quality. In general, the length of the GOP should take into account the video frame rate. A higher frame rate permits more frames to be used in a GOP. While MPEG is inherently variable bit rate (VBR) due to the different compression efficiencies for I, P, and B frames, the bits rates for the GOPs may exhibit smoother bit-rate variation, especially for low-motion content.

2.7.2 Closed GOP

In temporal prediction for a closed GOP, the first frame is an intracoded I frame and the last frame is a P frame. For all other frames in the GOP, P or B frames are used. Closed GOPs are self-contained since none of the frames refer to another frame outside the GOP. The frame pattern defines the periodic sequence of B and P frames in a GOP. Common frame patterns used in DVDs include IBP and IBBP. The frame pattern normally matches the GOP length. For example, an IBBP pattern may fit a GOP length of 10, giving a GOP sequence IBBPBBPBBP. However, an IBP pattern will lead to IBPBPBPBPB for a GOP length of 10, which is not valid for a closed GOP since the last frame is a B frame. A GOP length of 9 may be more appropriate, giving rise to IBPBPBPBP. Another possible sequence for a GOP length of 9 is IBBBPBBBP. This sequence uses one fewer P frame than the previous sequence, thus leading to slightly more efficient encoding than the previous frame pattern.

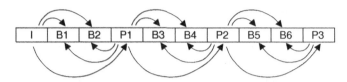

Figure 2.10   H.262/MPEG-2 temporal prediction for a closed GOP of length 10.

Figure 2.10 shows the H.262/MPEG-2 encoding process for temporal prediction using a closed GOP comprising 10 frames with an IBBP pattern. Previously coded I and P frames are usually employed as reference frames for B-frame interprediction and these I and P frames are always coded in display (picture) order. In this case, the I frame is always employed for forward prediction whereas the P frame can be used for forward or backward prediction. Note that the last P frame of a closed GOP only serves as a reference for two B frames instead of four. In addition, the I frame references up to three frames whereas a P frame references up to five frames. In general, the P frame references more frames than the I frame for closed GOPs that contain B frames (IBP, IBBP, IBBBP, etc.). As a consequence, I frames are buffered by the decoder for a shorter period of time than the P frames.

2.7.3   Error Resiliency in a Closed GOP

Clearly, if the I frame of a closed GOP is corrupted, all other frames within the GOP cannot be decoded. If an I frame is partially corrupted, the errors may propagate to other frames. In this case, the frames within the GOP may be decodable but visual artifacts may be introduced. Similarly, if any P frame is corrupted, the error may propagate to other P or B frames. B frames may be more vulnerable to error propagation (caused by previously decoded frames) than P frames since B frames require two reference frames as opposed to one reference frame for P frames. However, B frames may not contribute to propagation errors since they are usually not used as reference frames. Thus, corrupted B frames may be skipped by the decoder without perceptible degradation in the video quality. Using multiple reference frames may enhance the effort resiliency of the bitstream. However, employing more than two reference frames for the B frames may not further improve compression efficiency because the reference frames become increasingly outdated and bit overheads increase when more reference frames are included, especially for long GOPs.

Since each P frame requires one reference frame and each B frame requires two reference frames for interprediction, the encoder and decoder will need to store up to two reference frames. This is independent of the GOP length. The two reference frames for B-frame interprediction are either the I or P frames that precede or succeed the B frame in the display order. More specifically, an I frame and a P frame or two P frames are employed as reference frames for each B frame. Among all the B frames in a closed GOP, the first two B frames are the most accurately predicted since they refer to an intracoded I frame. They also require more cache memory since I frames are typically much bigger than P frames.

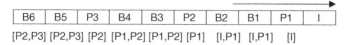

B6	B5	P3	B4	B3	P2	B2	B1	P1	I

[P2,P3] [P2,P3] [P2] [P1,P2] [P1,P2] [P1]   [I,P1]   [I,P1]   [I]

Figure 2.11   Decoded frame sequence (reference frames in parentheses).

2.7.4   Decoding Sequence

Figure 2.11 illustrates the frame-by-frame decoding sequence (which is also the transmitted frame sequence) for Figure 2.10. Clearly, the order of the video bitstream may not always correspond to the order of displaying the frames. For example, due to biprediction, B frames may be coded after but displayed before the previously coded I or P frame. The reference frames (e.g., I and P frames) must be decoded prior to the nonreference frames (e.g., B frames). In order to match the GOP length, a closed GOP may force a P frame at the end, thus disrupting the frame pattern. For example, when additional B and P frames are inserted at the end of an IBBP pattern with a GOP of length 12, this gives rise to the sequence IBBPBBPBBPBP. This process breaks the GOP pattern but allows the final B frame of the GOP to be decoded using two P frames as reference.

2.7.5   Open GOP

Figure 2.12 shows an example of an open GOP structure. A closed GOP with an IBBP pattern starts with an I frame whereas an open GOP with the same pattern may start with a B frame. Unlike the closed GOP, both I and P frames can be used for forward or backward prediction. In addition, the last P frame in a previous GOP is referenced by B frames in the current GOP. This GOP structure is commonly employed in Apple's HTTP live streaming (HLS). It ends with a P frame, just like a closed GOP. However, unlike a closed GOP, the open GOP fully exploits the last P frame, which is used as a reference for four B frames. As a consequence, fewer P frames may be employed when compared to closed GOP structures, giving rise to a slight improvement in compression efficiency. Note that the I frame now serves as a reference for more frames (5 frames), possibly as many as the P frame. Hence, interprediction is improved over the closed GOP and both I and P frames may be buffered by the decoder for the same period of time (i.e., a time interval corresponding to 5 frames).

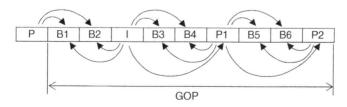

Figure 2.12   Open GOP with a length of 9.

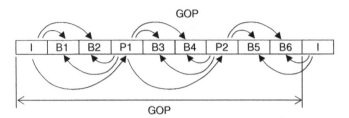

Figure 2.13   Open GOP with a length of 9.

For the same number of B frames in an IBBP GOP, two P frames are used for an open GOP compared to three in a closed GOP, giving rise to a smaller GOP length of 9 for the open GOP. The drawback of an open GOP is that it is no longer self-contained and hence, cannot be decoded independently. This will not apply to the first GOP of the video, which will start with an I frame. Alternative frame patterns of IBP and IBBBP confirm that an additional P frame can be omitted for the open GOP structure, thereby reducing its length by 1 compared to the closed GOP ([IBPBPBPBP] vs P[BIBPBPBP] and [IBBBPBBBP] vs P[BBBIBBBP]).

Another example of an open IBBP GOP structure is shown in Figure 2.13. Again, only two P frames are required for a GOP of length 9. This structure starts with an I frame, just like a closed GOP. In this case, the I frame is used as a reference for four B frames, including two from the previous GOP. Thus, the GOP need not end with a P frame. For the final GOP of the video, the last two B frames (i.e., B-5 and B-6) are not encoded.

2.7.6   Variable GOP Length

While defining a GOP length provides more structure to the video sequence, this is not always mandatory. A GOP with a variable length adheres to the frame pattern but allows the flexibility of inserting an I frame when the video content demands it. For example, when there is a new scene in the content, an I frame can be inserted. This may potentially lead to better compression efficiency than periodically inserting an I frame in a GOP. Video conferencing applications typically do not require an I frame for every group of 10 frames because the content is relatively static with few scene changes. By conserving the I frames, more B and P frames can be used to improve compression efficiency and the GOPs become longer. However, there is a limit on the maximum GOP length because the P frames are dependent on the I frames for referencing.

2.7.7   Random Access of MPEG Frames

An MPEG video bitstream may be accessed or switched at any time, such as during channel seeking (i.e., forward/reverse playback) and channel surfing. In many applications, the random access period typically varies from 1 to 10s. However, due to temporal prediction, a video decoder cannot start decoding a compressed video at a frame that is predicted from previous frames. The insertion of I frames allows random access to a video bitstream because it is encoded without any prediction from other frames. Thus, the latency in random access is inversely proportional to the rate of I frame insertion. The compression efficiency decreases as I frames are inserted more frequently because these frames only employ spatial compression. MPEG typically restricts the latency of random access of video frames at the receiver to 500 ms intervals. For a 30 Hz video, each frame is generated at 33 ms intervals, which implies 16 frames can fit into the 500 ms limit. Thus, the GOP length is limited to 16 frames for this case. More frames can be generated if a higher frame rate is used. This limit not only impacts interactive and cloud digital video recorder (DVR) applications due to the network propagation delay when users access the video from a remote location.

2.8 MOTION ESTIMATION AND COMPENSATION

Motion estimation is the key compression engine of many video coding standards. It exploits the similarity of successive frames (i.e., temporal redundancy) in video sequences. Many standards employ block-based motion estimation with adjustable block size and shape to search for temporal redundancy across frames in a video. When sufficient temporal correlation exists, MVs may be accurately predicted and only a small residual is transformed and quantized, thereby reducing the data needed to code the motion of each frame. Because objects tend to move between neighboring frames, detecting and compensating motion errors are essential for accurate prediction. Such techniques help partition and scale the bitstream with priority given to data that is more globally applicable. Thus, they not only improve the coding efficiency but also enhance error resilience.

2.8.1 Motion Estimation

Motion estimation or prediction requires each video frame to be partitioned into smaller blocks of samples. After temporal prediction, the MVs and reference frame indexes are entropy coded. A motion estimator attempts to determine the direction of minimum distortion for each block by matching similar areas or blocks from the preceding reference frame(s) to the corresponding areas in the current frame [4]. This recursive block matching process allows the best displacement or MV with the minimum distortion to be obtained for each block. The MV defines the magnitude and direction of the displaced areas. The probability of finding a matching block increases when smaller block sizes and more reference frames are used. If the blocks are small enough, the rotation and zooming of larger objects can be approximated by linear translation of these smaller areas. The MVs associated with each block can be differentially encoded with respect to adjacent blocks and may point outside the reference frame. The bit overheads for representing the MVs are further reduced using entropy coding. In addition, zero-valued MVs can be signaled efficiently without sending the MV.

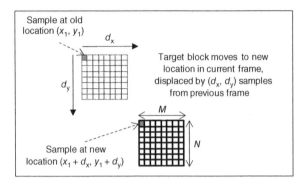

Figure 2.14   Motion estimation using linear translation.

2.8.2   Motion Search in P Frames

A P frame may predict motion by displacing the reference frame and finding matching blocks similar to the current coded block. Motion estimation using piecewise
linear translation is shown in Figure 2.14. The difference in displacement in the
horizontal and vertical directions with the best match that minimizes the error or
distortion is sent as an MV. Two popular error metrics, namely SSD and SAD, are
defined in equations (2.13) and (2.14), where B_k and B_{k-1} are blocks of $M \times N$ samples under comparison with the current frame (i.e., frame k) and previous frame (i.e.,
frame $k-1$), respectively. The maximum values for the horizontal (d_x) and vertical
(d_y) translations of the search window are usually ± 16 samples. The search is terminated if the SSD or SAD falls below an acceptable threshold, thus indicating that the
match is good. The accuracy of the search is generally not affected by the choice of
the error criterion although SAD is simpler to implement. One or more neighboring
blocks that have been coded previously can be used as an MV predictor for the current
block. Multiple MVs can be weighted (e.g., averaged) and a single MV is sent or the
median of the MVs can be sent. Alternatively, the current block can be subdivided
into smaller subblocks and a separate MV is transmitted for each of these subblocks.

$$\text{SSD}(d_x, d_y) = \sum_{j=1}^{N} \sum_{i=1}^{M} |B_k(x_i + d_x, y_j + d_y) - B_{k-1}(x_i, y_j)|^2 \qquad (2.13)$$

$$\text{SAD}(d_x, d_y) = \sum_{j=1}^{N} \sum_{i=1}^{M} |B_k(x_i + d_x, y_j + d_y) - B_{k-1}(x_i, y_j)| \qquad (2.14)$$

2.8.3   Motion Search in B Frames

B frames may predict the MVs using spatial prediction (based on neighboring blocks
from the current frame), temporal prediction (based on neighboring frames), or

weighted prediction (based on both reference frames). For P frames, MVs can be flagged as skipped if the motion characteristics of a block can be predicted from the motion of the neighboring blocks and the block contains no nonzero quantized transform coefficients. A similar mode (direct mode) exists for B frames. In this mode, the MV for a block is not sent but derived by scaling the MV of a colocated block in another reference frame or by spatially inferring motion from neighboring blocks. Note that skipped MVs are different from skipped frames with no knowledge of the motion trajectory of the samples. In this case, a skipped frame is either reproduced by repeating the preceding frame or by interpolation of the preceding and future frames. The former may result in jerkiness whereas the latter may lead to blurring of the moving areas.

2.8.4 Fractional (Subsample) Motion Search

The displacement of a block in a P frame relative to the reference frame may be specified by an integer MV or a fractional-precision MV (e.g., 1/2 or 1/4 sample precision), as shown in Figure 2.15. The motion prediction at these subsample locations is normally used for higher accuracy and often activated with RDO. In this case, the fractional samples are obtained by bilinear interpolation (or averaging) using a 1D multitap filter (e.g., 6-tap finite impulse response or FIR filter) applied in the horizontal and vertical directions. For example, in Figure 2.16, the 1/2 sample at location a can be obtained by averaging the integer sample values at horizontal locations A

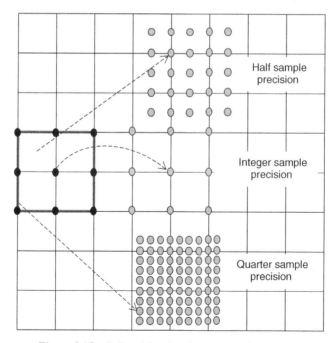

Figure 2.15 Full and fractional sample motion search.

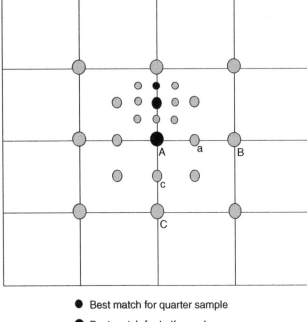

● Best match for quarter sample

● Best match for half sample

● Best match for integer sample

Figure 2.16   Interpolation in subsample motion search.

and B. Similarly, the 1/2 sample at location c can be obtained by averaging the integer sample values at vertical locations A and C. SSD or SAD can be used as the metrics to identify the best match for the predicted subsample MV.

A typical fractional sample motion search algorithm works as follows. The integer sample with the best match is first identified. That sample becomes the center of the 1/2 sample motion search. The values at 1/2 sample locations around the best match are bilinear interpolated using integer samples. The 1/2 sample with the best match, in turn, becomes the center of the 1/4 sample motion search. The values of the 1/4 sample locations are obtained by averaging samples at integer and/or 1/2 sample locations. The samples can be luma or chroma samples. Thus, by averaging the luma sample at integer and/or 1/2 sample locations, the sample at the 1/4 sample location is obtained. Chroma motion estimation uses color information for motion detection. The chroma sample at a fractional sample location is also obtained by averaging (horizontally, vertically, diagonally) via bilinear interpolation.

2.8.5   Motion Compensation

Motion compensation is necessary because motion prediction may not be perfect. For instance, when objects in a frame are rotated, a linear translation method for obtaining the MV may result in a residual. The residual is used to compensate for the MV and is normally subdivided into 8×8 blocks and sent to the block transform. The gain for motion compensation with integer-sample accuracy is typically less than 1 bit/sample. With subsample precision, the gain for motion compensation can be further improved. Motion compensation can also be improved using in-loop filtering.

2.8.6   Computational Complexity

Motion estimation is computationally intensive since the search may be performed at every sample location and in some instances, using multiple reference frames. Motion estimation algorithms generally rely on the assumption that most of the blocks in a frame are quasi-stationary or stationary. Hence, the search can be accelerated by using the recent frequently occurring MVs or the zero MV (corresponding to no movement) as a starting point. Figure 2.17 shows the directions of the MVs (depicted by arrows) for a hand moving in the downward direction. More changes in the MVs are associated with the index finger and the edges of the hand whereas many background MVs remain static.

Figure 2.18 illustrates a search window with integral projections of up to four samples using a 4×4 block with one-sample increments in eight different directions as depicted by the arrows. In this case, there is one reference block and the zero MV corresponds to the 4×4 block at the center of the grid. Including the zero MV, the total number of block comparisons is $1 + 8 + 8 + 8 + 8$ or 33. Each comparison requires 4×4 or 16 computations to obtain the SSD or SAD. Thus, the total number of computations is 16×33 or 528. However, many of these computations overlap with past search points and may be omitted. Alternatively, the search can begin at a larger step from the zero MV. For example, for high-motion content, one may start with a

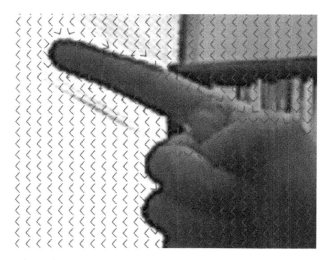

Figure 2.17   Directions of MVs for a hand moving downward.

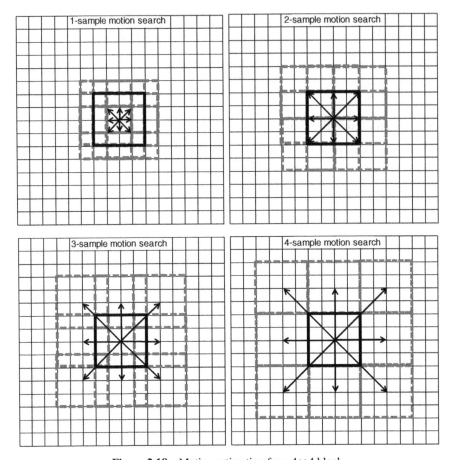

Figure 2.18   Motion estimation for a 4×4 block.

translation of four samples (instead of one sample) and then move the 4×4 block along the direction of the lowest SSD or SAD. In doing so, overlapped computations are avoided in the starting search, as can be seen in Figure 2.18.

To further reduce the number of block comparisons, the triangle inequality in equation (2.15) and the Cauchy–Schwarz inequality in equation (2.16) can be applied to equations (2.13) and (2.14), which can be rewritten as (2.17) and (2.18). In this case, the sums for the blocks in the current and previous frames are computed row-wise and column-wise before a single comparison between these sums is made. If the comparison indicates a quickly worsening SSD or SAD relative to a previous best match, individual block comparison is skipped. Such a strategy can significantly speed up an exhaustive full search algorithm [5]:

$$\sum_{i=1}^{M} |a_i| \geq \left| \sum_{i=1}^{M} a_i \right| \qquad (2.15)$$

$$\sum_{i=1}^{M} |a_i|^2 \geq \frac{1}{M} \left| \sum_{i=1}^{M} a_i \right|^2 \tag{2.16}$$

$$SSD(d_x, d_y) = \sum_{j=1}^{N} \sum_{i=1}^{M} |B_k(x_i + d_x, y_j + d_y) - B_{k-1}(x_i, y_j)|^2$$

$$\geq \frac{1}{NM} \left| \sum_{j=1}^{N} \sum_{i=1}^{M} B_k(x_i + d_x, y_j + d_y) - \sum_{j=1}^{N} \sum_{i=1}^{M} B_{k-1}(x_i, y_j) \right|^2 \tag{2.17}$$

$$SAD(d_x, d_y) = \sum_{j=1}^{N} \sum_{i=1}^{M} |B_k(x_i + d_x, y_j + d_y) - B_{k-1}(x_i, y_j)|$$

$$\geq \left| \sum_{j=1}^{N} \sum_{i=1}^{M} B_k(x_i + d_x, y_j + d_y) - \sum_{j=1}^{N} \sum_{i=1}^{M} B_{k-1}(x_i, y_j) \right| \tag{2.18}$$

2.8.7 Motion Search Algorithms

Motion estimation requires search algorithms to compute the MVs. An accurate search algorithm produces better video quality but may incur more time for coding. Conversely, a fast search algorithm may compromise video quality, especially for full-motion sports content. An intelligent search strategy can reduce computation while optimizing the video quality. The method of calculating the best predictor for the MV as well as the search area is usually vendor-specific. Some video codecs may apply intensity compensation to a reference frame before motion estimation. Several popular motion search algorithms are listed as follows:

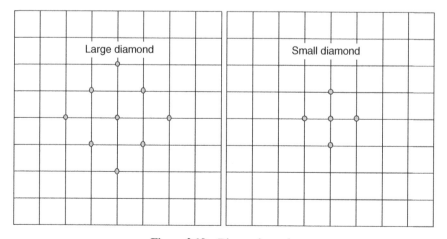

Figure 2.19 Diamond search.

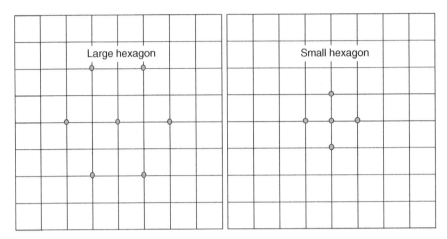

Figure 2.20   Hexagon search.

- *Diamond search*: A large diamond is first applied to the region of interest (Figure 2.19). If the best match (i.e., the location with minimum distortion) lies at the center of the large diamond, the large diamond is replaced by a small diamond with the same center and the search stops if the new minimum distortion location is identified from the small diamond. Otherwise, the center of the large diamond is moved to the location of the best match in the previous step and the search process is repeated. It is the fastest but may compromise video quality. Complexity is $O(n)$.

- *Hexagonal search*: This is a six-sided shape analysis method that provides reasonable quality at reasonable speed (Figure 2.20). It may employ fewer search locations compared to diamond search but may compromise video quality slightly. Like the diamond search, a large hexagon is replaced by a small hexagon when the best match is identified and the search stops after the new minimum distortion location is identified from the small hexagon. Otherwise, the center of the large hexagon is moved to the location of the best match in the previous step and the search process is repeated. Complexity is $O(n)$.

- *Uneven multihexagon search*: A more complex version of hexagonal search that produces high video quality at the expense of coding speed. Complexity is $O(n)$.

- *Full exhaustive search*: A complete and extensive analysis method where all possible displacements within the search region are examined. Although this brute-force method is slow and computationally expensive, the search pattern is highly regular, which may allow concurrent searches of multiple blocks of the same frame. A small improvement in video quality compared to the uneven multihexagon search method can be achieved. Complexity is $O(n^2)$.

- *Walsh Hadamard exhaustive search*: A more accurate version of exhaustive search method using Walsh Hadamard transform, which is also employed in pattern recognition. As shown in equation (2.19), the Walsh Hadamard transform H is the simplest orthogonal transform that can effectively eliminate spatial redundancies of an image. Since H is symmetric, it is equal to its transpose. However, this is the slowest among all search methods. It computes the sum of absolute transform differences (SATD) on each MV candidate, as shown in equation (2.20). Complexity is $O(n^2)$:

$$H = \begin{bmatrix} 1 & 1 & 1 & 1 \\ 1 & 1 & -1 & -1 \\ 1 & -1 & -1 & 1 \\ 1 & -1 & 1 & -1 \end{bmatrix} \tag{2.19}$$

$$\text{SATD}(d_x, d_y) = \frac{1}{2} \sum_{j=1}^{N} \sum_{i=1}^{M} |H[B_k(x_i + d_x, y_j + d_y) - B_{k-1}(x_i, y_j)]H^T|$$

$$= \frac{1}{2} \sum_{j=1}^{N} \sum_{i=1}^{M} |H[B_k(x_i + d_x, y_j + d_y) - B_{k-1}(x_i, y_j)]H| \tag{2.20}$$

Additional methods include subblock split and search, weighted prediction, weighted biprediction, and implicit weighted biprediction. Weighted prediction in P frames employs a predictor with scaling and offsets, and is especially effective for fade-in and fade-out scenes. The median MV is first obtained from the left, top, and top-left or right block. The difference between this median and the current MV is then entropy coded. For B frames, two MVs are allowed for each instance of temporal prediction. They can be obtained from any reference frame in the past or future. The weighted average of the sample values of the two reference frames are used as a predictor. The predictor can further employ scaling and offsets to apply the weighted average for biprediction, thereby mitigating cross-fade scenes.

2.8.8 Accelerating Motion Search

Motion estimation and intermode decision consume a large portion of encoder computation power and can be considered the most demanding portion of the encoding process. Motion estimation is the process of finding the MV for a block whereas intermode decision is the process of determining the best block size for interprediction. For motion estimation to adapt to the activity in the scene, smaller blocks can be used in detailed areas that are hard to compensate while larger blocks are appropriate when the motion or scene is homogeneous and highly correlated. In addition, the use of variable block sizes may provide better intraprediction since detailed areas of a frame can employ smaller block sizes for better resolution. To achieve the best coding efficiency, the encoder may test all possible modes and choose the best mode in terms of the least rate distortion or some Lagrangian cost function. Because variable block sizes may be used, this method will lead high computational complexity and limit the use of video encoders in real-time applications. In video coding, it is quite common that most of the transform coefficients in a block are quantized to zeros. In this case, a special symbol indicating this all-zero state is sent to the decoder instead of multiple zeros. Thus, motion estimation can be terminated early if an all-zero block is detected, thereby reducing unnecessary computations.

2.8.9 Impact of Video Resolution

For higher resolution videos, more samples should be used for motion estimation. A typical value of 16 is a good trade-off between speed and quality. Larger values can be used for more complex motion search. Consider a 720p video and 1080p video with the same content. If the number of MVs is identical for both videos, then inter-prediction is faster for the 720p video. However, the number of MVs cannot be the same because a higher resolution implies more blocks and thus more MVs. In addition, the magnitude of the MVs derived from videos with different resolutions may not be identical even if the trajectory of the motion is similar. For example, if a car is going from left to right, then many MVs will point in that direction independent of the resolution in the corresponding area of the frame.

2.9 NON-MPEG VIDEO CODING

Not all video codecs employ I, P, B frames or DCT as adopted by MPEG. In this section, we discuss the motion Joint Photographic Experts Group (JPEG) and Dirac video compression methods.

2.9.1 Motion JPEG

M-JPEG employs lossless intracoding for timed sequences of still images, which may be combined with audio. M-JPEG is widely used in Web browsers, media players, game consoles, digital cameras, streaming, and nonlinear (i.e., nondestructive) video production editing. It is based on discrete wavelet transform that works on the entire image (as opposed to blocks). This achieves good compression efficiency without exploiting temporal redundancy. However, newer video coding standards such as H.265/HEVC may achieve superior compression efficiency than M-JPEG even if only intracoded I frames are employed. M-JPEG combines context models with arithmetic codes for entropy encoding. Because the wavelet transform is a generalization of the Fourier transform, it is computationally more intensive. However, since each frame is independently encoded or decoded, M-JPEG is computationally faster than MPEG, which requires complex motion estimation and compensation of interdependent frames. As such, the delay in MPEG encoding can be a few orders of magnitude higher than M-JPEG. The difference is small for decoding.

Due to M-JPEG's accurate postcompression rate control, the coded bits per frame are almost constant. Rate control in MPEG compression is based on modeling and feedback. When combined with temporal interframe compression, MPEG may lead to a high variation in the number of coded bits per frame, thereby requiring more buffering that increases coding delay. In terms of the compression efficiency, MPEG with interprediction works best on low to moderate motion videos whereas M-JPEG works best on high-motion videos. M-JPEG exhibits high resiliency to compression and transmission errors. More specifically, M-JPEG tends to experience only short duration artifacts when individual frames are corrupted. In contrast, errors in MPEG references frames may propagate over subsequent interpredicted frames until resynchronization (refresh) occurs on the next I frame. Due to I frame refresh, interframe coding, and rate control, MPEG video quality may vary over time and motion artifacts may be more visible.

2.9.2 Dirac

Dirac is an open-source royalty-free video coding technology that employs a 2D 4×4 discrete wavelet transform to remove spatial redundancies [6]. The transform allows Dirac to operate on a wide range of video resolutions because entire frames, as opposed to smaller blocks, are used. Like block-based transforms such as DCT, it decorrelates data in a frequency-sensitive way but preserves fine details better and thus, achieves high coding efficiency. Dirac provides 1/8 sample MV precision and uses arithmetic coding but not in-loop filters. Dirac supports CBR and VBR operation as well as lossy and lossless compression. The lossless mode can be used for high-quality and low-latency applications (e.g., professional production).

2.9.3 WebM Project

VP8 is an open-source video format formerly owned by Google but released as part of the WebM project [7], which was launched in May 2010. VP8's data format and decoding process are described in RFC 6386 [8]. VP8 was designed to be more compact, easy to decode, and royalty-free. By adopting a freely available Web platform, VP8 targets faster innovation and better user experience. VP9 is the latest open video codec that became available on June 17, 2013. An overview of the VP9 bitstream is described in [9]. VP9 is supported by Web browsers that understand HTML5, including Chrome and Firefox. Android does not natively support WebM well. Google announced on January 11, 2011 that future versions of its Chrome browser will no longer support H.264/AVC [10]. Both VP8 and VP8 employ nonadaptive entropy coding, which leads to faster encoding and more consistent coding gains.

2.10 CONSTANT AND VARIABLE BIT-RATE VIDEOS

Compressed videos can be broadly categorized under CBR and VBR. In general, CBR encoding may result in variable video quality due to rate caps imposed by the encoder on a group of encoded frames according to the desired output bit rate. The caps may be chosen based on network requirements. For example, a compressed CBR video can be capped to about 1.5 Mbit/s to fit into a T1 line rate of 1.536 Mbit/s.

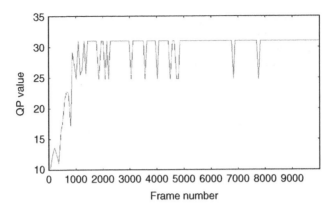

Figure 2.21   QP variation with CBR coding.

2.10.1   CBR Encoding

In CBR encoding, the maximum bit rate is capped to a user-specified value. The video quality assumes more importance. To this end, the encoder dynamically adapts the bit rate of each frame according to previously coded frames and limits the variations of the bit rate in subsequent frames so that perceptible disruptions in the video quality are minimized. However, in some situations, CBR encoding may result in temporary degradation in the video quality. For instance, frames with a large size (due to a scene change or high motion) may be encoded using a higher QP value in order to maintain a constant coded bit rate at the output. The variation in the QP values for a CBR video is shown in Figure 2.21.

RDO via a rate controller can be applied to minimize degradation of future frames in the CBR encoding (see Chapter 3 for more details). This may increase encoder complexity. The minimum encoding delay corresponds to the length of the GOP period. It may not be possible to measure the overall video quality of a CBR video since this is variable and may change for each frame due to changes in the QP value and content complexity. The variation is more noticeable with HD videos as shown in Figure 2.22. For some video content, the bit rate may remain variable even with CBR encoding. This is because RDO may attempt to maintain a minimum video quality level and as a result, may increase the bit rate beyond the rate limit to compensate.

2.10.2   VBR Encoding

VBR encoding is favored for its superior compression efficiency, which makes the coded videos easier to store and stream. VBR encoding is an open loop scheme that maintains the video quality at a constant level because each video frame is coded using a fixed QP. This simplifies the codec design, and minimizes encoding and decoding time. Each frame can be encoded with a delay that is shorter than the frame duration, thus facilitating real-time services. In this case, the compression efficiency assumes more importance. Because there are no rate caps, the video bit rate can be

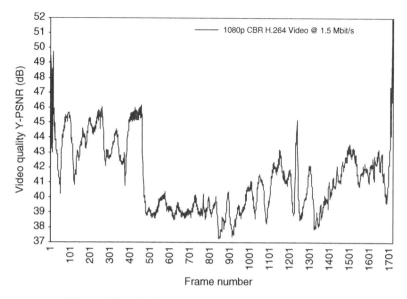

Figure 2.22   Variable video quality with CBR HD coding.

bursty and content dependent. For example, scene changes or high motion tend to generate more bits per frame than "talking heads" or video conferencing applications with little motion.

Statistically multiplexing several VBR video streams can help smooth the bit-rate variation of the aggregated output. This allows the number of video streams to be increased within a fixed channel bandwidth by exploiting self-averaging variations in the instantaneous VBR bit rates. For example, if one stream is demanding high rate, it is likely other streams have capacity to spare. Although additional bit and delay overheads are required in identifying and separating the streams at the receiver, the delay can be lower than per stream buffer-based smoothing.

2.10.3   Assessing Bit Rate Variability

Streaming VBR video traffic is a special challenge due to the high dynamic range of the frame sizes that results in high bit-rate variability. The compressed frame sizes of a 1080p VBR video are shown in Figure 2.23. Some frames exhibit larger sizes compared to the average. The peak values are primary caused by scene changes, which are normally intracoded. High motion may also lead to large frame sizes but the peak values are typically lower due to motion prediction. For the highest peak value, the peak to average ratio (PAR) of the frame size can approach 30. When videos with the same content but different resolutions are coded with the same QP, the PARs of the coded frame sizes for both videos are almost identical even though the actual frame sizes for the 1080p video are larger than the 720p video (Figure 2.24). Smoothing the VBR video using memory buffers may reduce the bit-rate variability (Figure 2.25).

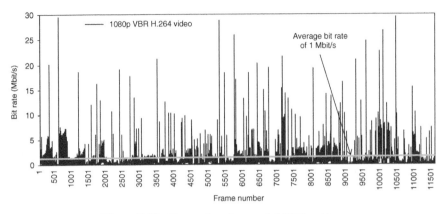

Figure 2.23   Frame sizes for a coded 1080p video.

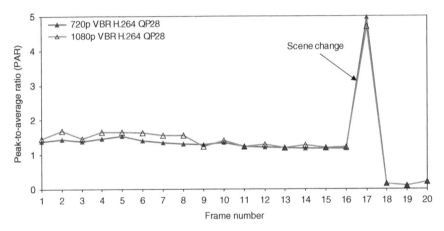

Figure 2.24   PAR of coded frame sizes for 720p and 1080p videos (same content).

The smoothed video makes efficient use of available bandwidth and reduces peak rates that may cause packet losses over the network or buffer overflow at the receiver.

2.10.4   Scene Change Detection

A typical consequence of a scene change is a significant change in the frame size, which can be larger or smaller than the previous scene. A scene change can be classified as sudden and gradual. Abrupt changes are easy to detect as two successive frames may be uncorrelated. This implies the encoded frame sizes of the frames can be very different and can be used as a metric to detect abrupt scene changes. Gradual changes are used to enhance the quality of video production and are more difficult to detect as the difference between the frames corresponding to two successive scenes is small, thus leading to only small changes in the encoded frame sizes.

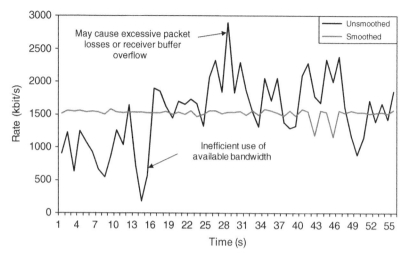

Figure 2.25   VBR video smoothing.

Some encoders provide a frame size threshold for scene change detection so that an intracoded frame can be inserted at every scene change (instead of an intercoded frame), resulting in better looking scene transitions. Better video quality for high-action movies, sports, and trailers can be achieved in this way when intracoded frames are inserted more frequently and selectively. Since a reliable scene change detector is required to determine the specific time instants to insert the intracoded frames, this in turn increases the encoding delay. The threshold of the scene change detector determines its effectiveness. A high threshold reduces the impact because the encoder will detect fewer scene changes, which tend to be abrupt scene changes. A low threshold yields more aggressive detection and the encoder starts over more often. This may prevent real-time video encoding but enables detection of both gradual and abrupt scene changes.

2.10.5   Adaptive Scene Change Detection

The potential correlation of two successive frames during a scene change occurrence suggests that we can use the SAD to adapt the scene change threshold to the video. The threshold $T(n)$ for frame n can be selected using equation (2.21). X_{n-1} is the encoded size of frame $n - 1$. X_{n-1} can also be replaced by X_{n+1}:

$$T(n) = aX_{n-1} + bm_n + c\sigma_n \qquad (2.21)$$

where $m_n = \frac{1}{N} \sum_{i=n-N-1}^{n-1} X_i$ and $\sigma_n = \sqrt{\frac{1}{N-1} \sum_{i=n-N-1}^{n-1} (X_i - m_n)^2}$

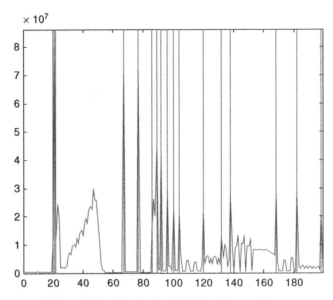

Figure 2.26   Correlation of SAD and scene change.

Figure 2.26 shows the case when a threshold with $a = 0$, $b = 1$, and $c = 2.5$ is applied to the first 200 frames of the high-action video trailer. The vertical lines indicate the actual occurrences of a scene change. There were 14 correctly detected scene changes, 3 missed, and 2 false positives, thus giving a hit rate of 74%. Since this video contains very fast scene changes, the performance of the scheme should be better for most videos. The detection method can be modified to include changes in the video quality between successive frames in a scene change.

2.10.6   I Frame Size Prediction

Intracoded I frames generally take up the most bandwidth because its size is much larger compared to the other frame types. It turns out that the size of successive I frames can be highly correlated. Figure 2.27 shows the autocorrelation of 7500 I frames for a 60-min movie encoded using H.264/AVC with an IPBBPBBPBB GOP structure. A lightweight algorithm that makes use of the size of two previous I frames can be employed to predict the size of the next I frame. As can be seen in Figure 2.28, the prediction is highly accurate. A similar idea can be used to estimate the size of P and B frames as well as a GOP.

2.11   ADVANCED AUDIO CODING

Advanced audio coding (AAC) is the standard audio format for smartphones, tablets, game consoles, YouTube, and broadcast TV standards. Like MPEG video, AAC is a

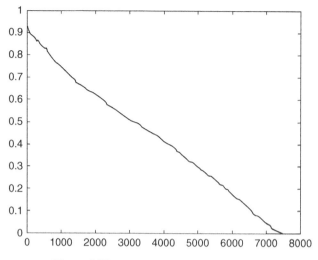

Figure 2.27 Autocorrelation of I frame size.

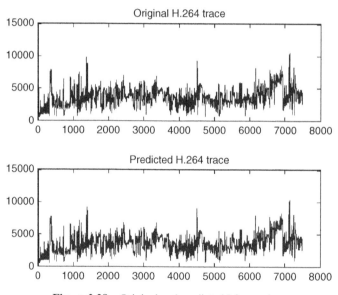

Figure 2.28 Original and predicted I frame size.

lossy compression scheme for coding digital audio and is standardized as MPEG-4 Part 3 (MP4 audio). It is a successor to MP3 and AC3 (Dolby Digital), and generally achieves better sound quality at comparable bit rates. AAC supports three profiles and employs new techniques such as perceptual noise substitution, temporal noise shaping, backward adaptive linear prediction, and enhanced joint stereo coding. It supports a broad range of sampling rates (8–96 kHz), bit rates (16–576 kbit/s), and audio channels (1–48).

2.11.1 Low and High Bit Rate AAC

Two types of AAC are adopted for MPEG video transport. In low bit-rate AAC, the transport of one or more complete AAC frames of variable length is supported. The maximum length of an AAC frame in this mode is 63 bytes. For each AAC frame contained in the payload, the one byte AU-header contains information on the length of each AAC frame in the payload and the index for computing the sequence timing of each AAC frame. High bit-rate AAC also supports AAC frames with variable lengths. However, the maximum length of an AAC frame in this mode is 8191 bytes. For each AAC frame contained in the payload, the AU-header provides the same information as the low rate case. Unlike the low rate case, each 2-byte AU-header is used to code the maximum length of an AAC frame (13 bits) and the AU-index (3 bits). For stereo coding at 64 kbit/s, each high bit-rate AAC frame contains 200 bytes on the average, allowing seven frames to be carried over a 1500-byte Ethernet payload.

2.11.2 High-Efficiency and Low-Complexity AAC

While AAC is optimized for high quality use, high-efficiency advanced audio coding (HE-AAC) is designed for low bit-rate streaming (typically less than 128 kbit/s). Low-complexity advanced audio coding (LC-AAC) is a subset of HE-AAC. For LC-AAC, prediction is not used and there is a lower order temporal noise shaping filter. Apple's HLS allows stereo audio of up to 48 kHz to be encoded in HE-AAC or LC-AAC. An AAC stereo audio stream with a 48 kHz bandwidth typically produces a bit rate of 160 kbit/s. HE-AAC and AAC coders are available [11].

2.11.3 MPEG Surround

MPEG Surround (MPS) is an audio compression technique for multichannel audio signals [12]. It is standardized as MPEG-D Part 1. When combined with HE-AAC, MPS can carry a 5- or 7-channel surround program at scalable data rates, usually 64 kbit/s or less. These bass channels are sometimes accompanied by a low-frequency effects (LFE) channel that requires only one-tenth of the bandwidth of the main audio channels (a LFE channel is sometimes known as a subwoofer channel), giving rise to 5.1 or 7.1 surround systems.

2.12 VIDEO CONTAINERS

Video is usually combined with audio into a single file. Not only will this allow the video and audio to sync up in local playback (such as using a DVD player), it is also important in enabling a video program to be delivered over an IP network. Typically this involves combining a few elementary streams, which may comprise one or more video/audio channels and optional data channels (e.g., subtitles, closed captions). An elementary stream contains only one type of data (e.g., audio, video, program information). The elementary stream can also be packetized, giving rise to a packetized elementary stream (PES). The PES header contains timing information such as a common clock base to indicate when the video bitstream can be decoded and presented. The elementary stream can be encapsulated in containers such as MPEG-4 (.mp4), AVI (.avi), QuickTime (.mov), Flash video (.flv), RealMedia (.rm), Matroska (.mkv), and MPEG-2 program stream (.mpg) or transport stream (.ts). MP4 is a popular container for HTML5 browsers. FLV is a popular container for non-HTML5 compliant browsers. Video advertisements also commonly employ the FLV format to maximize compatibility.

2.12.1   MPEG-4

The MPEG-4 (MP4) Part 14 (ISO/IEC 14496-14) file format is a multimedia container specified as a part of MPEG-4 [13]. MP4 specifies the encapsulation of audio-visual information in an audio or video elementary stream. The MP4 media object contains audio and video, which can be presented in an audio-visual scene as well as streaming data. Streaming data in the MP4 media objects may employ one or more elementary streams. An object descriptor is used to identify all streams associated with a single media object and allow handling of coded content and meta information associated with the content. The streams may be file-based or URL-based applications. Each individual stream may be further characterized by a set of descriptors containing encoder/decoder configurations and hints to the quality of service (QoS) needed for transmission (e.g., maximum bit rate, bit error rate, priority, etc.). Synchronization of the elementary streams is achieved via time stamping of individual access units (AUs) contained within the streams.

2.12.2   MP4 Access Units

Information on the type of MPEG-4 stream carried is conveyed by the multipurpose Internet mail extension (MIME) format parameters. Each MPEG-4 elementary stream consists of a sequence of AUs, which may include compressed audio and video data. The AU is the smallest data entity that contains timing information. For audio, the AU may represent an audio frame and for video, a video frame. All AUs are byte-aligned and zero bits are padded for frames that are not byte-aligned. AUs can be interleaved to improve error resiliency.

2.12.3   Binary Format for Scenes

The binary format for scenes (BIFS) is the framework for the MP4 presentation engine. The BIFS system defines the spatio-temporal arrangements of the audio/video objects in the scene. Structures marked as "template" in the ISO base media file format (BMFF) pertain to the composition, including fields such as matrices, layers, graphics modes (and their colors), volumes, and balance values. A box is the elementary syntax element in the BMFF, including the type, byte count, and payload. A BMFF file consists of a sequence of boxes and boxes may encapsulate other boxes. A movie ("moov") box contains the metadata for the continuous media streams that

Figure 2.29   Two BMFF media tracks with fragments.

are present in the file. Each stream is represented as a track, which is denoted as track ("trak") box. The media content of a track is either enclosed in a media data box ("mdat") or directly in a separate file. The media content for the tracks consists of a sequence of samples, such as audio or video AUs. The BMFF specifies the following types of tracks:

- Media track containing an elementary media stream;
- Hint track comprising either media transmission instructions or a received packet stream;
- Timed metadata track comprising time synchronized metadata.

Although originally designed for storage, the BMFF is very useful for adaptive streaming. For example, the movie fragments defined in BMFF can be used. Figure 2.29 shows a fragmented BMFF file with two tracks, which may contain video and audio. After reception of the "moov" box, any movie fragment ("moof") with its associated media data can be decoded.

2.12.4   MP4 Overheads

The overhead for encapsulating a H.264/AVC video in MP4 is insignificant, typically below 0.01%. The MP4 container incurs lower overheads than the MPEG-2 TS container, which may contain unnecessary padding. The difference between MP4 and MPEG-2 TS overheads is in the region of 4%.

2.12.5   MPEG-2 TS

MPEG-2 TS is a packet-oriented container that defines the multiplexing and synchronization of compressed video and audio data or AUs. Because MPEG-2 TS is widely deployed in cable, satellite, digital TV broadcast, and HLS systems as well as optical disk storage, it is sometimes used to carry H.264/AVC media streams.

2.12.6   MPEG-2 TS Structure

An MPEG-2 TS consists of a series of fixed 188-byte transport stream packets (TSPs). Each TSP typically requires 4 bytes of video header or 7 bytes of audio header. The start of the TSP contains a synchronization byte of 0x47. This is followed by the payload unit start indicator (1 bit), adaptation field control (1 bit), packet identifier

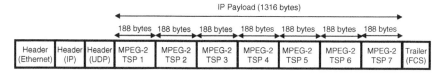

Figure 2.30   MPEG-2 TS encapsulation in Ethernet/IP.

(10 bits), continuity counter (4 bits), and payload byte offset (8 bits). The payload of each TSP may contain PES data. The packet identifier is used to identify a PES. Synchronization in MPEG-2 TS is based on the system target decoder (STD) model, which is used by video codecs to ensure interoperability. The STD specifies the buffer size to store the coded media data and the time when an AU can be removed from the buffer and decoded. It is very similar to the hypothetical reference decoder (HRD) in H.264/AVC and H.265/HEVC.

2.12.7   MPEG-2 TS Audio and Video PESs

The audio and video PESs are sequentially separated into TSPs and each TSP contains information from one elementary stream only. The PES header starts from the first byte of the TSP payload. The length of the PES header is variable: 8 bytes for video without B frames and audio, 13 bytes for video with B frames. An 8-byte PES header contains the start code prefix (3 bytes, 0x000001), the stream identifier (1 byte), the PES length (2 bytes, which gives a maximum PES length of 65,536 bytes), and the timestamp for synchronization purposes (2 bytes). The length of the PES payload is also variable since the length of the audio and video streams are variable.

2.12.8   MPEG-2 TS IP/Ethernet Encapsulation

As shown in Figure 2.30, up to seven 188 byte TSPs can be encapsulated in an IP packet. The minimum overhead required to carry 7 TSPs over IP is about 3.4%. This is because IP/Ethernet encapsulation adds 46 bytes of overhead per packet. With 7 TSPs occupying 188×7 or 1316 bytes, the percentage overhead becomes $46/1362 \times 100\%$ or 3.4%.

2.13   CLOSED CAPTIONS

Closed captions display text on a video to provide additional or interpretive information. The audio portion of the video is transcribed, sometimes with nonspeech descriptions. The captions are not visible until activated by the viewer. The FCC requires all content previously aired on TV to include captions when distributed online. Popular formats include SubRip (SRT), scenarist closed caption (SCC), and Web video text tracks (WebVTT) [14], the emerging HTML5 standard. The VTT image format allows individual scenes to be previewed with thumbnails at specific points in the video. Legacy CEA 608 closed caption data can be converted to WebVTT.

REFERENCES

1. J. Han, A. Saxena, V. Melkote, and K. Rose, "Jointly Optimized Spatial Prediction and Block Transform for Video and Image Coding," IEEE Transactions on Image Processing, Vol. 21, No. 4, April 2012, pp. 1874–1884.

2. J. Han, Y. Xu, and D. Mukherjee, "A Butterfly Structured Design of the Hybrid Transform Coding Scheme," IEEE Picture Coding Symposium, 2013.

3. FFMPEG, http://www.ffmpeg.org.

4. J. R. Jain and A. K. Jain, "Displacement Measurement and Its Application in Interframe Image Coding," IEEE Transactions on Communications, Vol. 29, No. 12, December 1981, pp. 1799–1808.

5. Y. C. Lin and S. C. Tai, "Fast Full-Search Block-Matching Algorithm for Motion-Compensated Video Compression," IEEE Transactions on Communications, Vol. 45, No. 5, May 1997, pp. 527–531.

6. Dirac, http://diracvideo.org.

7. WebM Project, http://www.webmproject.org.

8. VP8 Data Format and Decoding Guide, January 2013, https://tools.ietf.org/html/rfc6386.

9. A VP9 Bitstream Overview, February 2013, https://tools.ietf.org/html/draft-grange-vp9-bitstream-00.

10. HTML Video Codec Support in Chrome, http://blog.chromium.org/2011/01/html-video-codec-support-in-chrome.html.

11. AAC Coding, http://www.audiocoding.com.

12. MPEG Surround, http://www.mpegsurround.com.

13. ISO/IEC 14496-14 International Standard, "Information Technology – Coding of Audio-Visual Objects – Part 14: MP4 File Format," November 2003.

14. WebVTT, http://dev.w3.org/html5/webvtt.

HOMEWORK PROBLEMS

2.1. Show that a raw 60-min 480p (4:4:4) video at 30 Hz requires roughly 112 Gbytes of storage. How many minutes of a raw UHD-1 video (4:4:4) at 30 Hz will require the same storage requirements? A DVD has a storage capacity of 4.7 Gbytes. Compute the number of DVDs needed to record 90 min of the a raw UHD-1 video (4:4:4) at 30 Hz. Estimate the total bit rate when the DVDs are delivered to a customer in under 30 min using unmanned aerial vehicles (e.g., Amazon's Prime Air). If H.265/HEVC is used to compress the UHD-1 video, compute the compression efficiency needed to store the video in one DVD.

2.2. Show that a 720p video frame comprises 3600 MBs.

2.3. Show that a 1080p (4:4:4) video at 60 Hz is roughly equivalent to a raw bit rate of 3 Gbit/s. What is the percentage reduction in the bit rate when the video is converted to black and white? Repeat the calculations for a 4:2:0 video.

2.4. Explain why an integer number of 16×16 MBs cannot be supported by a 1080p video.

2.5. HDMI 2.0 (UHD HDMI) increases the maximum digital signaling rate per channel from 3.4 Gbit/s (in HDMI 1.4) to 6 Gbit/s. Show that HDMI 2.0 may support UHD-1 at 60 Hz using a single channel.

2.6. Suppose each I frame of a compressed 480p video requires an average of 30 kbytes for transmission. The P frame requires an average of 3 kbytes and no B frames are employed. The GOP structure employs one I frame and five P frames and the frame rate is 30 Hz. Determine the appropriate data rate of the network to transport this video in bits per second. If 100 kbytes of buffering is available at the transmitter, compute the minimum data rate.

2.7. For older TVs, the Y component is derived from equation (2.22), where $0.299 + 0.587 + 0.114 = 1$.

$$Y = 0.299R + 0.587G + 0.114B \tag{2.22}$$

Why is a different formula used from equation (2.1)? For black and white TVs, are all RGB components needed? The YC_bC_r values can be further scaled as follows:

$$Y' = 16 + 219Y \tag{2.23}$$

$$C_b = 128 + 224[0.5/(1 - 0.114)](B - Y) \tag{2.24}$$

$$C_r = 128 + 224[0.5/(1 - 0.299)](R - Y) \tag{2.25}$$

Using these equations, confirm the Y', C_b, C_r values for Table 2.10.
The sample values may employ shifted 8-bit values with a scaling factor of $2^{(n-8)}$, where n is 10 or 16. For example, the 8-bit value for white is 235, which gives a 10-bit value of 235×4 or 940. The 10-bit format may also use 16 bits for each Y, U, V component, with six of the least significant bits set to zero.

Table 2.10 $Y'C_bC_r$ Values.

Color	R	G	B	Y'	C_b	C_r
Black	0	0	0	16	128	128
Red	255	0	0	81	90	240
Green	0	255	0	145	54	34
Blue	0	0	255	41	240	110
Cyan	0	255	255	170	166	16
Magenta	255	0	255	106	202	222
Yellow	255	255	0	210	16	146
White	255	255	255	235	128	128

2.8. Explain whether large GOP sizes (i.e., I frame followed by many P or B frames before the next I frame) are appropriate for the following content: (a) action movies, (b) newscasts, (c) sports, (d) documentaries, and (e) video chats.

2.9. What are the disadvantages of breaking the pattern in a closed GOP?

2.10. Compare the pros and cons of simple video compression methods as listed:

- Remove pixels at regular intervals, essentially scaling the image or making it more blocky;
- Changing the pixel aspect ratio by subsampling – taking several adjacent pixel values and averaging them together, resulting in a single rectangular pixel that approximates the value of several;
- Remove color channel information at regular intervals (e.g., 4:2:2, 4:2:0).

2.11. For the bitstream 01010101010101010101010101010101010101010101010101, determine the minimum number of transmitted bits using lossless encoding. Is such encoding effective for audio compression?

2.12. The mismatch in progressive frame rate between film (typically 24 Hz) and television (typically 25/30 Hz) creates a need for "pulldown" methods that stretch the frame rate by exploiting the nature of 60 Hz interlaced (also called 60i) videos. Besides frame repetition and insertion of null frames, devise other pulldown methods to achieve this objective. It is common that 30 and 24 Hz videos are played back at 29.97 Hz (i.e., 30/1.001 Hz) and 24.98 Hz (i.e., 24/1.001 Hz), respectively. How can synchronization problems be avoided for these cases?

2.13. Compare the raw bit rates for a 720p video versus a 1080i video. Assume that 60 Hz videos are used (60p or 60i). How will the rates compare to a 1080i 60 Hz video that is deinterlaced before being transmitted? Compare the rates for a 1080i video versus a 1080p video at 60 Hz. What can you conclude about the bandwidth requirements? Why do some ATSC and pay-TV channels prefer 720p over 1080i? Will 720p provide better video quality than 1080i when playing high-motion sports videos?

2.14. In many HDTVs, it is common to have several display modes, for example, wide zoom, zoom, normal, full, and so forth. Suppose you set the HDTV to 1080p with a display aspect ratio of 16:9. Explain whether this ratio will change when different display modes are selected. Assuming square pixels, compute the display aspect ratio for 1080i and compare with 720p and 1080p. Explain why most computer display resolutions are 4 by 3 (e.g., 640×480, 1024×768) and 8 by 5 (e.g., 1280×800, 1440×900). How will a 1080p video appear on a 720p HDTV and a 1280×800 computer display? Will the pixel to aspect ratio and the display to aspect ratio change?

2.15. High refresh rate HDTVs such as 240 Hz HDTVs may employ interpolation between frames to create and insert additional frames into a lower frame rate video. To convert a 60p video to 240 Hz, how many additional frames must be inserted in 1s? Many movie projectors operate at 24 Hz. However, each frame is illuminated two or three times before the next frame is projected using a shutter in front of its lamp. As a result, the projector has a 48 or 72 Hz refresh rate. Can this method be adopted to display a 24p video (commonly used in Blu-ray players) on a 240 Hz HDTV?

2.16. The DVB asynchronous serial interface (DVB-ASI) is similar to serial digital interface (SDI) and frequently used in satellite transmission. A DVB-ASI program stream utilizes 7 MPEG-2 TS packets per IP packet. One stereo audio pair at 256 kbit/s is included in the stream. Together with a 480i H.264/AVC video (60 Hz) that is coded using the 4:2:2 color format with 0.25 bit/pixel, compute the bit rate of the transport stream. How many program streams can be multiplexed in a 35 Mbit/s multiprogram transport stream (MPTS)?

2.17. The YUV 4:1:1 color format employs $4Y$ samples for every $1C_b$ and $1C_r$. In this case, 4:1 horizontal subsampling is performed but there is no vertical subsampling. Derive the total number of bits per pixel if 8 bits/luma sample is used.

2.18. Redo the arithmetic coding example with $P(a) = 1/3$ and $P(b) = 2/3$.

2.19. Suppose an old black and white video image contains specks of noise. Will DCT or ADST provide better coding efficiency when the image is compressed?

2.20. If an intracoded I frame is inserted at every scene change, is there a possibility of intracoded P and B frames?

2.21. JPEG 2000 employs wavelet transforms and codes regions of interest (ROI) using several mechanisms that support spatial random access or ROI access at varying degrees of granularity. In this way, it is possible to store different parts of the same picture using different quality. JPEG 2000 is robust to bit errors due to the coding of data in relatively small independent blocks. The Portable Network Graphics (PNG) format employs the RGB color format and achieves superior compression for images with many pixels of the same color, such as computer-generated graphics. Explain whether each of these image compression methods can be applied to intracoding in video compression.

2.22. Prove that the number of $N \times N$ block comparisons for all projections within a search window of $(2W + N) \times (2W + N)$ where W is the maximum displacement is $(2W + 1)^2$. What are the advantages of using the diamond shape versus the square shape in motion search?

2.23. For a video sequence consisting of M frames encoded with a given quantization level, if X_n ($n = 1, 2, \ldots, M$) denotes the size of an encoded video frame, then the covariance (CoV) of the encoded video is defined as (2.26). In this case, σ

is the standard deviation and $\overline{X}$ is the mean of the frame sizes:

$$
\text{CoV} = \frac{\sigma}{\overline{X}} = \frac{\sqrt{\frac{1}{(M-1)} \sum_{n=1}^{M} \left(X_n - \frac{1}{M} \sum_{n=1}^{M} X_n \right)^2}}{\frac{1}{M} \sum_{n=1}^{M} X_n} \tag{2.26}
$$

Derive a relationship between the CoV and the PAR of the frame sizes. Compute the CoV using the PAR for 10 frames of sizes 20, 10, 5, 5, 5, 5, 5, 20, 5, 20 kbytes. Confirm your answer by computing CoV directly using (2.26). Now consider CBR frame sizes 10, 10, 10, 10, 10, 10, 10, 10, 10, 10 kbytes, which give the same average as the previous case. Compute the CoV and the PAR.

3

H.264/AVC STANDARD

This chapter covers several important aspects of H.264/MPEG-4 advanced video coding (AVC) video coding standard. H.264 currently dominates Web, mobile, and connected TV sectors. Virtually all new smartphones and tablets support H.264 using the MP4 container, including Apple, Android, and Windows Mobile. H.264 is also a key video format for HTML5, which is supported by major browsers such as Chrome, Firefox, Internet Explorer, and Safari. H.264 overcomes many limitations of motion estimation in H.262/MPEG-2 with improved interprediction via fine-grained motion estimation, multiple reference frames, unrestricted motion search, and motion vector prediction. H.264 also offers improved intracoding in the spatial domain using a context-sensitive deblocking filter and many techniques for mitigating errors, packet losses, and bandwidth variability, such as the use of scalable bitstreams (corresponding to different quantization levels) and slice coding (with no possibility of spatial error propagation from one slice to any other slice within the video frame). The improved coding efficiency and the new error resilient features make this standard a perfect candidate for video streaming, broadcasting, and conferencing over a variety of fixed and mobile networks.

3.1 OVERVIEW OF H.264

H.264 is a digital video coding standard developed by the Joint Video Team (JVT) and standardized by the International Telecommunications Union (ITU) [1]. The standard

Next-Generation Video Coding and Streaming, First Edition. Benny Bing.
© 2015 John Wiley & Sons, Inc. Published 2015 by John Wiley & Sons, Inc.

is also known as AVC or MPEG-4 Part 10. H.264 promises to deliver the same quality video while consuming up to 50% lower transmission bandwidth or storage file size when compared to H.262. It can also be used for sending higher quality/resolution videos for a given bit rate. H.264 provides bit rate adaptivity and scalability, two key features that address the needs of different applications when operated over heterogeneous networks. This is necessary when the receiving device is not capable of displaying full resolution or full quality images. H.264 allows new extensions to be added as technology improves. The Joint Model (JM) reference software [2] provides a test platform for normative features of the H.264 standard. It is not optimized for commercial use due to the long encoding and decoding delays. Test sequences and bitstreams are also available.

3.1.1 Fundamental H.264 Benefits

H.264 offers high coding efficiency, lower storage requirements, and reduces visual artifacts. H.264 provides scalable source coding for multiple resolutions. The bit rate and resolution of the content can be adapted to networking environment and display device. For the same content, H.264 typically achieves a twofold improvement in coding efficiency over H.262 and hence provides higher quality video using the same bandwidth or the same quality video using lower bandwidth. A higher coding efficiency can be achieved for HD than SD videos, allowing a large amount of content to be stored on a single optical disc.

3.1.2 H.264 Applications

H.264 is widely adopted in consumer electronics equipment such as camcorders, surveillance cameras, tablets, and smartphones. H.264 is the recommended codec for all digital video broadcast (DVB) and 3GPP video services and is adaptable to different applications, client devices, and networks. For instance, the video quality may be prioritized over coding efficiency when bandwidth is abundant and is useful for applications when consumer device is capable of displaying the full resolution or full quality video. Because H.264 decoders can be implemented readily in software, this helps jumpstart online TV services, allowing over-the-top (OTT) providers to achieve quality TV delivery that is comparable to pay-TV services. Cable TV service is still very much dependent on legacy H.262 set-top boxes (STBs) that were extensively deployed over a decade ago. The cost for upgrading millions of these STBs to H.264 may be prohibitive. However, H.264 STBs have become indispensable for enabling new IPTV services over DSL.

3.2 H.264 SYNTAX AND SEMANTICS

Like prior MPEG standards, H.264 defines the syntax and semantics of the bitstream (i.e., the coded video representation) and the processing that the decoder requires. It does not define how encoding and other video preprocessing functions are performed, including coding optimization, thus enabling vendors to differentiate their encoders in terms of cost, coding efficiency, error resilience, or hardware requirements. In the H.264 standard, the term "slice" is preferred over "frames" or "pictures" when describing the different prediction types. Thus, I, P, and B slices are often used.

3.2.1 Profiles and Levels

H.264 profiles and levels specify restrictions on the bitstreams and hence limit the capabilities needed to decode the bitstreams. Each profile specifies a subset of algorithmic features and limits that shall be supported by all decoders conforming to that profile. Three basic profiles (baseline, extended, main) are defined (Table 3.1). The high profiles are called fidelity range extensions (FRExt) and may achieve further coding efficiency than H.262, as much as 3:1. They include all main profile coding tools that contribute to the coding efficiency for the 8-bit/sample 4:2:0 sampling format. H.264 also defines 16 levels together with the profiles (Table 3.2). The levels specify upper limits for the frame size (in MBs) ranging from QCIF to 4K UHD, decoder processing rate (in MB/s) ranging from 250 Kpixels/s to 250 Mpixels/s, video bit rate ranging from 64 kbit/s to 240 Mbit/s, and video buffer size. To decode the H.264 video, both the video profile and level must be supported. Note that a lossless MB prediction mode can be enabled in the High 4:4:4 profile. In this mode, sample values are interpredicted but sent without transformation (i.e., transform bypass). In H.262, only four profiles (simple, main, high, 4:2:2) and four levels are defined.

3.2.2 Baseline, Extended, Main Profiles

The baseline profile is a subset of the extended profile. Both profiles address video transmission in error-prone environments. The baseline profile minimizes complexity at the expense of the lowest coding efficiency. The extended profile supports more error resilience techniques and is the only profile that supports SI and SP slices (see Section 3.10.2). It reduces temporal correlation using B slices but is computationally more complex. However, the extended profile is not a popular choice in deployment. The baseline profile is used when short coding and decoding times are desired (e.g., video conferencing, video chat applications). For this reason, B slices are not allowed in the baseline profile. The main profile provides the best video quality among the basic profiles at the expense of higher decoder complexity, primarily due to the use of B slices and context-adaptive binary arithmetic coding (CABAC).

3.2.3 High Profiles

The high profiles employ more than 8 bits/sample and allow higher resolution sampling formats (i.e., 4:2:2 and 4:4:4). The higher profiles target higher quality videos using more chroma samples per luma sample (e.g., 4:4:4 vs 4:2:0) or finer QP (up to 12 bits/sample for high 4:4:4). Thus, the use of higher profiles is justified if the video is already in good quality. For each additional bit of source video accuracy, the quantization step increases by 6. A decoder conforming to the high 4:4:4 profile will be

Table 3.1 H.264 Profiles and Capabilities.

Parameter	Baseline	Main	Extended	High	High 10	High 4:2:2	High 4:4:4
Use of I and P slices	X	X	X	X	X	X	X
Use of B slices		X	X	X	X	X	X
Use of SI and SP slices			X				
Interlaced coding (PicAFF, MBAFF)		X	X	X	X	X	X
Data partitioning			X				
Redundant slices	X		X				
Flexible MB ordering	X		X				
Arbitrary slice ordering	X		X				
Multiple slice groups	X		X				
8-bit sample depth	X	X	X	X	X	X	X
9- to 10-bit sample depth					X	X	X
11- to 12-bit sample depth							X
Transform bypass							X
Quantization scaling matrices				X	X	X	X
Weighted prediction for P and SP slices		X	X	X	X	X	X
CABAC		X		X	X	X	X
CAVLC	X	X	X	X	X	X	X
8×8 transform decoding				X	X	X	X
Predictive lossless coding							X
Separate picture scaling				X	X	X	X
Separate C_b and C_r QP control				X	X	X	X
Residual color transform							X
Monochrome format				X	X	X	X
4:2:0 Sampling format	X	X	X	X	X	X	X
4:2:2 Sampling format						X	X
4:4:4 Sampling format							X

Table 3.2 H.264 Levels.

Level Number	Maximum MBs per second	Maximum Frame Size (MBs)	Maximum Video Bit Rate (VCL) for Baseline, Extended, and Main Profiles (Mbit/s)	Maximum Video Bit Rate (VCL) for High Profile (Mbit/s)	Maximum Video Bit Rate (VCL) for High 10 Profile (Mbit/s)	Maximum Video Bit Rate (VCL) for High 4:2:2 and 4:4:4 Predictive Profiles (kbit/s)	Examples of Maximum Resolution @ Frame Rate (Maximum Stored Frames) in Level
1	1,485	99	0.064	0.080	0.192	0.256	128×96 @ 30.9 (8) 176×144 @ 15.0 (4)
1b	1,485	99	0.128	0.16	0.384	0.512	128×96 @ 30.9 (8) 176×144 @ 15.0 (4)
1.1	3,000	396	0.192	0.24	0.576	0.768	176×144 @ 30.3 (9) 320×240 @ 10.0 (3) 352×288 @ 7.5 (2)
1.2	6,000	396	0.384	0.48	1.152	1.536	320×240 @ 20.0 (7) 352×288 @ 15.2 (6)
1.3	11,880	396	0.768	0.96	2.304	3.072	320×240 @ 36.0 (7) 352×288 @ 30.0 (6)
2	11,880	396	2	2.5	6	8	320×240 @ 36.0 (7) 352×288 @ 30.0 (6)
2.1	19,800	792	4	5	12	16	352×480 @ 30.0 (7) 352×576 @ 25.0 (6)
2.2	20,250	1,620	4	5	12	16	352×480 @ 30.7 (10) 352×576 @ 25.6 (7) 720×480 @ 15.0 (6) 720×576 @ 12.5 (5)
3	40,500	1,620	10	12.5	30	40	352×480 @ 61.4 (12) 352×576 @ 51.1 (10) 720×480 @ 30.0 (6) 720×576 @ 25.0 (5)

(continued)

Table 3.2 (*Continued*)

Level Number	Maximum MBs per second	Maximum Frame Size (MBs)	Maximum Video Bit Rate (VCL) for Baseline, Extended, and Main Profiles (Mbit/s)	Maximum Video Bit Rate (VCL) for High Profile (Mbit/s)	Maximum Video Bit Rate (VCL) for High 10 Profile (Mbit/s)	Maximum Video Bit Rate (VCL) for High 4:2:2 and 4:4:4 Predictive Profiles (kbit/s)	Examples of Maximum Resolution @ Frame Rate (Maximum Stored Frames) in Level
3.1	108,000	3,600	14	14	42	56	720×480 @ 80.0 (13) 720×576 @ 66.7 (11) 1280×720 @ 30.0 (5)
3.2	216,000	5,120	20	25	60	80	1280×720@ 60.0 (5) 1280×1024@ 42.2 (4)
4	245,760	8,192	20	25	60	80	1280×720 @ 68.3 (9) 1920×1080 @ 30.1 (4) 2048×1024 @ 30.0 (4)
4.1	245,760	8,192	50	62.5	150	200	1280×720 @ 68.3 (9) 1920×1080 @ 30.1 (4) 2048×1024 @ 30.0 (4)
4.2	522,240	8,704	50	62.5	150	200	1920×1080 @ 64.0 (4) 2048×1080 @ 60.0 (4)
5	589,824	22,080	135	168.75	405	540	1920×1080@72.3 (13) 2048×1024@72.0 (13) 2048×1080@67.8 (12) 2560×1920@30.7 (5) 3680×1536@26.7 (5)
5.1	983,040	36,864	240	300	720	960	1920×1080@120.5 (16) 4096×2048@30.0 (5) 4096×2304@26.7 (5)

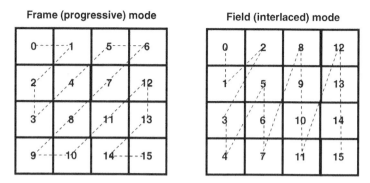

Figure 3.1   Frame and field scan order for 4 × 4 blocks.

capable of decoding a bitstream coded with high 4:2:2, high 10, high, or main profiles. Similarly, the high 4:2:2 profile decoder is capable of decoding the high 10, high, and main profiles. These profiles do not offer tools for error robustness or resilience and are mainly designed for storage or for broadcasting. They achieve higher coding efficiency using weighted prediction for P slices, 8 × 8 luma prediction with low-pass filtering to improve predictor performance, and adaptive transform coding for the intracoded slices.

In adaptive transform coding, 4 × 4 blocks may be used for more detailed parts of the frame while 8 × 8 blocks are applied to areas with less details. Adaptive transform coding is supported by perceptual-based quantization scaling matrices that take into account the visibility of a specific frequency associated with each transform coefficient to tune the quantization fidelity. The quantized coefficients are then scanned in the zigzag fashion before being entropy encoded. The scan ordering is designed to order the highest-variance coefficients first and to maximize the number of consecutive zero-valued coefficients appearing in the scan. Figure 3.1 shows H.264 scan order in the frame and field modes for 4 × 4 blocks.

3.3   H.264 ENCODER

A H.264 encoder is shown in Figure 3.2. The common modules with the decoder are highlighted in bold. The decoder conceptually works in reverse, comprising the entropy decoder and the common modules.

3.3.1   H.264 Slice Types

H.264 defines three slice types:

- Spatially compressed intracoded instantaneous decoding refresh (IDR) I slices (can be decoded independently with no reference to other slices);
- Unidirectionally predicted P slices (predicted from previous I or P reference slices);

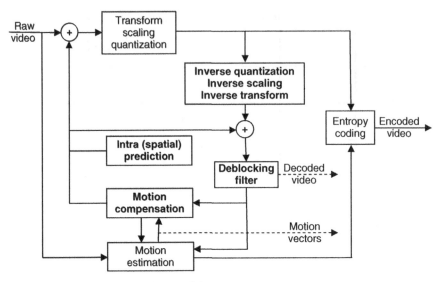

Figure 3.2   H.264 encoder.

- Bidirectionally predicted B slices (predicted from previous and future reference slices).

An I frame normally contains multiple I slices, whereas a P or B frame is normally contained in one P slice or one B slice, respectively. A closed group of pictures (GOP) starts with an IDR slice, which allows all subsequent slices to make reference to it. A B slice can serve as a reference slice. This is limited by the maximum decoded frame buffer size and the video resolution.

3.3.2   H.264 Intraprediction

Intracoded slices enable error resilience and random access to the bitstream but achieve moderate coding efficiency. The intracoded blocks in each slice employ spatial prediction from previously coded samples within the same slice to minimize spatial correlation. The difference between the predicted and coded block is encoded. The key objective is to remove predictable low frequency components, which can be perfectly reconstructed by the decoder. This is achieved using directional linear interpolation of adjacent edge blocks from neighboring blocks that are coded before the current block. These neighboring blocks are to the left and/or above the current block to be predicted (H.262 only allows prediction from the left block). An encoded parameter identifies the neighbors and provides information on how they should be used for prediction.

If the selected neighboring blocks are intercoded, error propagation due to motion compensation may occur if these blocks become corrupted. A constrained intracoding mode can be signaled that allows prediction only from intracoded neighboring blocks. Because of the relatively small areas, significant spatial correlation exists with each directional mode. If there is no directional pattern, the DC predictor is employed. A smaller partition always gives equal or better compression than a larger one when the extra overhead is ignored. Overall, there are four luma prediction modes for 16×16 MBs, nine luma prediction modes for 4×4 blocks, and four chroma prediction modes for 8×8 blocks.

3.3.3 Intraprediction for 4×4 Blocks

The eight spatial prediction directions for 4×4 intraprediction are shown in Figure 3.3. The DC mode (mode 2) is nondirectional. Among these nine modes, the vertical (mode 0), horizontal (mode 1), and DC modes are frequently used. The remaining modes are diagonal modes. The modes 0, 1, 3, and 4 are shown in Figure 3.4. A single line with an arrowhead indicates that the samples along the line are copied from the previously coded sample. For a line with a double arrowhead, the values of two previously coded samples are averaged. A maximum of 13 boundary samples (A–L and Z) from previously coded blocks are employed. If samples E–H are not available (e.g., they may lie outside a slice), these samples are replaced by sample D. Mode 0 uses samples from the top boundary for prediction. Similarly, mode 1 employs samples from the left boundary for prediction. The DC luma predictor requires both the top and left boundaries of the previously coded blocks. For this mode, all 16 samples of the 4×4 block share the same prediction, which is the weighted average of eight boundary samples A–D and I–L (sample Z is excluded). The encoder then follows up each prediction sequentially by applying the discrete cosine transform (DCT) in vertical and horizontal directions. Note that modes 0 and 1 require boundary samples from only one previously coded block. The DC mode and mode 8 require two whereas modes 3–7 require three blocks.

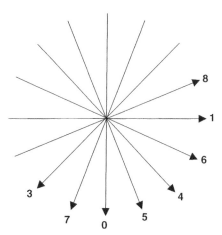

Figure 3.3 Intra (spatial) predictor directions for 4×4 blocks.

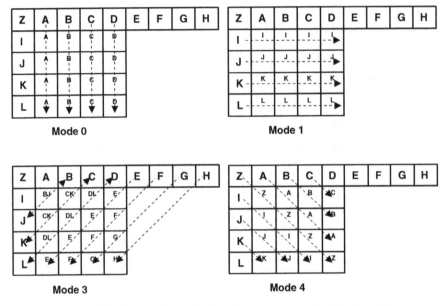

Figure 3.4   Intraprediction for 4×4 blocks using modes 0, 1, 3, and 4.

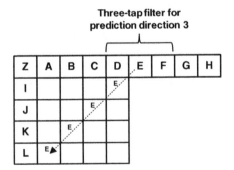

Figure 3.5   Multitap filtering in intraprediction.

A multitap filter can be applied to any of the prediction directions, as shown in Figure 3.5.

3.3.4   Intraprediction for 16×16 Macroblocks

The four intraprediction modes for 16×16 MB prediction are vertical, horizontal, DC, and plane. The vertical and horizontal modes are illustrated in Figure 3.6. The plane mode employs a linear function between the neighboring samples to the top and left coded MBs to predict the current sample. Similarly, for 4:2:0 chroma intraprediction, the vertical, horizontal, DC, and plane modes are used and prediction is performed on 16×16 MBs since chroma is usually smooth over large areas.

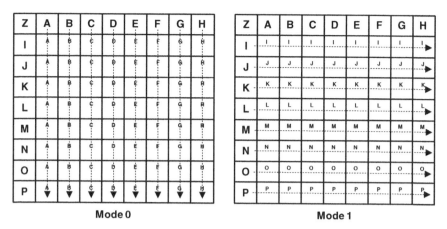

Figure 3.6 Intraprediction for 16×16 MBs using modes 0 and 1.

3.3.5 Intra Pulse Code Modulation Mode

Intra pulse code modulation (IPCM) coding allows the encoder to simply bypass intraprediction and transform coding processes and directly send the values of the coded samples. The IPCM mode allows the encoder to accurately represent the sample values of unusual video content without significant overheads. It also imposes an upper limit on the number of bits a decoder must handle for an MB without degrading coding efficiency.

3.3.6 H.264 Interprediction

Unlike intrapredicted slices, interpredicted slices employ prediction from previously decoded reference slices and motion compensation to reduce temporal correlation among the slices. The prediction error is transformed, quantized, and entropy encoded, and transmitted together with the prediction mode information. At the decoder, the scaled and quantized transform coefficients are inverse transformed and added to the predicted signal. The minimum luma motion-compensated block size is 4×4. Bipredicted slices may also be used as references for coding other slices.

Besides intracoded MBs, there are MBs that can be partitioned into various block shapes for motion-compensated prediction. Luma P and B block sizes of 16×16, 16×8, 8×16, and 8×8 samples are supported. 8×8 samples may be further split into 8×4, 4×8, or 4×4 samples. These partitions are illustrated in Figure 3.7. If a single intercoded MB is split into four 8×8 blocks and each 8×8 block is further split into four 4×4 partitions, then 16 MVs together with the index for the reference frame is transmitted. The smaller blocks improve the ability to handle fine motion detail and result in better subjective viewing quality because they do not produce large blocking artifacts.

Motion-compensated prediction requires both encoder and decoder to store the reference frames, which are signaled using an index. The reference index parameter

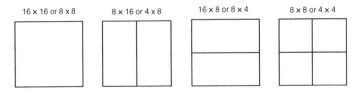

16 × 16 or 8 × 8   8 × 16 or 4 × 8   16 × 8 or 8 × 4   8 × 8 or 4 × 4

Figure 3.7   Partitions for interprediction.

is transmitted for each luma P block. Subblocks smaller than 8×8 use the same reference index. In P-skip coding, reference index 0 is employed and no quantized prediction error, MV, or reference index is transmitted. This is useful for coding smooth areas or large areas with constant motion such as slow panning. B-coded blocks employ two reference lists: list 0 and list 1, which are managed by the codec's buffer control. The four supported predictions include list 0, list 1, biprediction, and direct prediction. Biprediction is achieved by a weighted average of two distinct motion-compensated prediction values from list 0 and list 1. Unlike prior standards where two predicted values are averaged, in H.264, arbitrary weights are allowed. In addition, it is possible to use B slices as reference images for further predictions. Direct prediction is inferred from a previous prediction (i.e., list 0, list 1, or biprediction). Like the P blocks, the B blocks can also be coded in the B-skip mode.

The accuracy of motion compensation is fine-grained and may vary from full, 1/2, and 1/4 of the luma sample size. If an MV points to an integer-sample location, the prediction signal consists of the corresponding samples of the reference slice; otherwise the corresponding sample is obtained using interpolation to generate fractional locations. The prediction values at 1/2 sample locations are obtained by applying a 1D six-tap finite impulse response (FIR) filter horizontally and vertically. Prediction values at 1/4 sample locations are generated by averaging samples at integer and 1/2 sample locations. The predicted values for the chroma component are obtained by bilinear interpolation. Since the sampling resolution of chroma is lower than luma, 1/8 sample accuracy is employed for chroma locations.

3.4   RATE DISTORTION OPTIMIZATION

The basic goal of lossy coding is to provide good rate distortion. H.264 MBs can be intracoded (in the spatial domain) or intercoded (in the temporal domain). Depending on the intracoding mode, the luma component of each MB can be uniformly predicted or subdivided into smaller 4×4 blocks for prediction. The smaller blocks are useful for coding highly detailed areas of the frames, including edges and uneven regions. In both cases, several prediction techniques are available. The intercoding modes may also subdivide the MBs for more accurate motion-compensated prediction.

Rate distortion optimization (RDO) minimizes both distortion and rate using Lagrangian optimization techniques. The best prediction mode is chosen by jointly minimizing the distortion and the number of bits for coding the block or MB. Both parameters are dependent on the chosen QP and the prediction mode. The distortion can measured as the squared or absolute difference between the original block and the reconstructed block. There are two flavors of RDO, namely, fast high complexity and high complexity. The fast high complexity mode uses a simplified algorithm with fewer computations and thus, reduces the encoding time. However, this is done at the expense of the video quality.

3.4.1 RDO under VBR

The H.264 video compression efficiency generally becomes poorer when RDO is activated. However, the impact of RDO on the size of the encoded video is dependent on the video content. With high-quality video coding (i.e., when a low QP is used), the encoding time increases significantly with RDO compared to the case without RDO but the video quality improvement is greater than a high QP. This improvement is achieved on each coded slice and is independent of the specific temporal sequences in the video such as the GOP structure or the use of B slices. To sum up, RDO in the variable bit rate (VBR) mode provides better video quality for higher quality videos but the encoded video is typically larger in size and the encoding time is longer. Thus, RDO is well suited for broadcasting of prerecorded high-quality videos but should be avoided for low delay applications.

3.4.2 RDO under CBR

This mode determines if the desired video quality can be maintained when the size of the encoded H.264 video (i.e., the video bit rate) remains fixed. RDO simply employs the best prediction mode depending on the complexity of the current scene in the video content. However, before employing a specific QP value, the rate control algorithm requires information that is only available after the RDO algorithm has completed its prior computations. In fact, to simplify the coding process, the impact of RDO is predicted based on the complexity of the previous scenes in the video sequence. The encoding time remains relatively constant as the bit rate increases without RDO whereas it increases steadily with the bit rate when RDO is used. Because better video quality can be achieved with a higher bit rate, the QP decreases, resulting in a longer encoding time. A summary of the key RDO performance metrics is depicted in Table 3.3.

Table 3.3   RDO (High Complexity) When Compared to No RDO.

Mode	Encoding Time	Compression Efficiency	Video Quality
VBR	Longer (especially for low QP)	Lower	Better for every frame (especially for low QP)
CBR	Longer	NA	Better (especially for low QP)

3.4.3 In-Loop Deblocking Filter

Deblocking is a computationally intensive postprocessing step (performed after MB decoding) that adaptively smoothes the edges between adjacent blocks, thereby reducing blockiness caused by block processing and quantization. It is particularly useful for lower quality videos, which is more susceptible to blocky artifacts. The deblocked image is buffered and used to predict the motion of future frames in a coding loop. This improves the video quality because past reference frames used for motion compensation will be filtered versions of the reconstructed frames. The effectiveness of the adaptive filter can be adjusted to produce a sharper video or remove more details. The threshold for blockiness detection can also be adjusted and this is dependent on the quantizer. For example, when the quantization step size is small, the effect of the filter is reduced. The filter is turned off when the quantization step size becomes very small.

3.5 VIDEO CODING AND NETWORK ABSTRACTION LAYERS

The hierarchical H.264 architecture is shown in Figure 3.8. H.264 employs two conceptually different layers to adapt the video bitstream to different transport networks:

- Video coding layer (VCL);
- Network abstraction layer (NAL).

3.5.1 Video Coding Layer

The VCL defines the core video coding functions such as block partitioning, intraprediction, motion-compensated prediction, transform coding, entropy coding, and in-loop filtering. These low level functions are transport unaware and are designed to efficiently represent the video content. The highest data structure for VCL is the video frame, which comprises an integer number of MBs typically arranged in raster scan order. The NAL is an interface between the VCL and the transport network. It is, therefore, responsible for encapsulation of the coded video data into NAL units, a format that facilitates video transmission or storage.

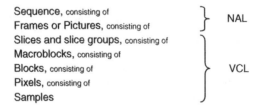

Figure 3.8   Hierarchical H.264 architecture.

3.5.2 Network Abstraction Layer

An NAL unit is a coded video fragment of variable length with header information that enables packetized information to be transmitted across the network. It comprises a 1-byte header (that indicates the payload type) and a bit string that is formed by the video data payload. NonVCL information such as sequence and picture parameter sets (PPSs), access unit (AU) delimiters, supplemental enhancement information (SEI) or video usability information (VUI) may also be added. The parameter set structure provides a robust mechanism for conveying data that are essential to the decoding process. The sequence parameter set (SPS) applies to a sequence of consecutive coded video frames whereas the PPS applies to the decoding of one or more individual frames within a coded video sequence. The SPS and PPS decouple the transmission of infrequently changing information from the transmission of coded video frames. NAL units are separated by a 4-byte flag (0x00000001). The Real-time Transport Protocol (RTP) payload supports three NAL modes:

- *Single mode*: Each NAL unit is transported in a single RTP packet;
- *Noninterleaved mode*: Several NAL units of the same frame are packetized into a single RTP packet;
- *Interleaved mode*: Several NAL units from different frames are packetized into a single RTP packet, not necessarily in their decoding order.

A set of NAL units is called an AU. The decoding of each AU results in one video frame. SEI (e.g., frame timing information) may also precede the coded video frame. A coded video sequence may contain IDR AUs, each containing an intracoded frame. Subsequent AUs in the bitstream will not reference AUs prior to the IDR AU.

3.5.3 Hypothetical Reference Decoder

The hypothetical reference decoder (HRD) specifies constraints on the coded NAL unit or bytestream such as how bits are fed to a decoder and how the decoded frames are removed from a decoder. It verifies the conditions of the video buffer to ensure decoder compatibility among hardware devices. Buffer fullness is checked to ensure no overflow or underflow. It also specifies the maximum buffer size and the initial buffer occupancy at the start of playback. The HRD is similar to the video buffer verifier (VBV) in H.262/MPEG-2, which determines the output bit rate of CBR streams.

Two buffers are specified namely the coded picture buffer (CPB) and the decoded picture buffer (DPB). The CPB models the arrival and removal times of the coded bits. The hypothetical stream scheduler (HSS) is a delivery mechanism that provides timing and data flow information to input a bitstream into the HRD. It is used to check for conformance of a bitstream or a decoder. H.264 employs a sliding window mechanism to mark the reference frames in the DPB in order to derive which reference frames are to be kept in the DPB for interprediction.

3.5.4 Supplemental Enhancement Information

Just like metadata may be included in video containers to allow efficient search for stored H.264 videos, SEI can also be added on a frame-by-frame basis to allow the container to store "in band" data. SEI can be used for a variety of purposes. It is typically coded in a proprietary format since there is no standardized format for unregistered user data in H.264. By providing the current frame number, it can be used for stream positioning. It may indicate the stream bit rate or depth information for 3D video decoding. For example, stereo video SEI allows the coder to identify the use of the video (e.g., left and right views) on stereoscopic displays. Another use of SEI is to provide decryption information such as the decryption key or the seed for deriving the decryption key for the encrypted video stream. SEI can be employed to validate the frame using a checksum or hash-based message authentication code (HMAC). The newly created SEI NAL unit may itself be encrypted and/or signed (validated) so that information contained in it is not easily accessible to an unauthorized user and thus, the video content can be used for security purposes without being compromised.

3.6 ERROR RESILIENCE

H.264 offers a number of error resilience methods, which improve the reliability in compressed video delivery. These methods typically require overhead bits to be added by the encoder. The NAL ensures a consistent video syntax to be used in different network environments. The SPS and PPS provide more robustness and flexibility than prior designs. The SPS and PPS, together with the IDR frame, enable a bitstream to be decoded from any random point. If errors are unavoidable, frame numbering allows H.264 to detect and localize frame losses. It also allows the creation of subsequences and enables temporal scalability by optional inclusion of extra frames between other frames. More powerful error resilience methods spread the errors at the receiver, allowing error concealment of the damaged areas to perform effectively. This minimizes the visual impact of losses on the actual distorted image. Contrast the use of interleaving at the packet or bit level, which help spread burst errors such that forward error correction (FEC) methods can perform more effectively. Two popular methods are data partitioning (DP) and flexible macroblock ordering (FMO).

3.6.1 Slice Coding

H.264 employs slices to support parallel encoding/decoding and also offers error resiliency. A slice corresponds to a contiguous group of MBs in a video frame (Figure 3.9). One or more slices are needed to cover the entire frame. A frame may contain slices of different types and may be used as a reference for interprediction of subsequent slices regardless of its slice coding type. This generalization enables prediction structures such as hierarchical B slices that improve the coding efficiency compared to the IBBP GOP structure typically used in H.262.

The actual number of MBs within a slice is not standardized. In H.264, the minimum length of a slice can be less than a row of MBs corresponding to the length of

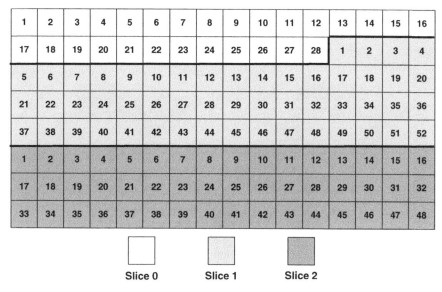

Figure 3.9   H.264 slice coding.

the frame. This is clearly more flexible than restricting the minimum length of each slice to a single row of MBs in the frame. Thus, H.264 slices need not be of rectangular shape. Granularity can be a single MB if desired. It is a requirement that if the frame is subsequently used for predictions, then predictions shall only be made from those regions of the frame that are enclosed in slices. However, some information from other slices may be needed for deblocking filtering across slice boundaries.

With slice coding, each frame is subdivided into slices and the order can be modified when an error resilience method is used. The slice is given increased importance because it is the basic spatial segment. Each slice is independently coded from its neighbors. In this way, errors or missing data from one slice cannot propagate to any other slice within a frame. This in turn, increases the flexibility to extend frame types (I, P, B) down to the level of slice types. Redundant slices are also permitted. Bits below and above the frame layer are related to the VCL and NAL, respectively.

3.6.2   Data Partitioning

Data partitioning separates compressed video information into different levels of importance. Each level is encapsulated in a single data packet and is, therefore, independently decodable. Higher-priority syntax elements (e.g., sequence headers) can be separated from lower-priority data (e.g., B slice transform coefficients). Thus, transmission of each level can be prioritized or accorded unequal protection, which may in turn minimize loss rates for important data and improve robustness.

Each video frame can be divided into three partitions namely A, B, and C. Partition A assumes the highest importance since it carries critical parts of the bitstream such

Partitions B and C drop for I, P, B frames   **Partition B and C drop for B frames**

Figure 3.10   Dropping partitions for different frame types.

as control and sequence header information, DC coefficients, QP values, and MVs. Thus, partition A data should not be dropped or left unprotected during transmission, otherwise partitions B and C become useless. Partition B contains intracoding information while partition C (biggest partition) contains intercoding information, including less critical data such as the higher frequency DCT coefficients. Since intracoding information can prevent error propagation, partition B is more important than partition C. This means more frequent partition B updates may be needed, leading to larger B partitions.

Higher-priority partitions (e.g., partition A) in DP can be protected with stronger FEC to ensure that all users, including multicast users, receive the video at the minimum quality with no artifacts. Lower-priority partitions (e.g., partitions B and C) can be accorded with standard FEC to reduce the network overheads. This is an alternative to designing a video server that adapts its transmission to suit many individual receivers simultaneously. Figure 3.10 shows the impact of dropping partitions for different frame types.

3.6.3   Slice Groups

H.264 allows the possibility of dividing a video frame into regions or partitions called slice groups (SGs). The partitioning is specified by the MB to SG map, which identifies the slice group of each MB in the frame. This information is included in the PPS and the slice headers. Each SG may comprise one or several slices. Since each slice represents a sequence of consecutive MBs (processed in raster scan order – left to right and top to bottom) within the same unique SG, a slice can be decoded independently. However, the MBs in an SG are not necessarily in raster scan order with respect to a frame. For example, in Figure 3.11, the second row of MBs for SG 1 starts from the third MB rather than the first MB of that row. H.264 limits the number of SGs in a frame to eight to prevent complex allocation schemes.

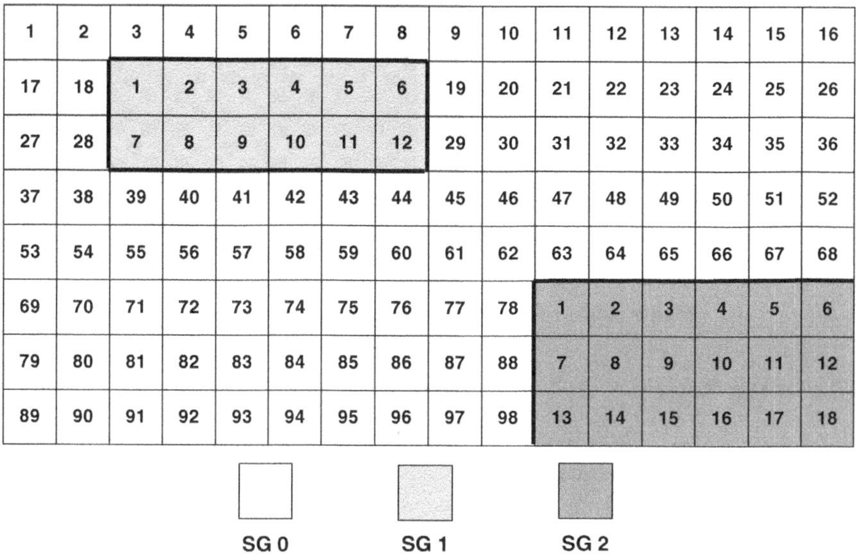

1	2	3	4	5	6	7	8	9	10	11	12	13	14	15	16
17	18	1	2	3	4	5	6	19	20	21	22	23	24	25	26
27	28	7	8	9	10	11	12	29	30	31	32	33	34	35	36
37	38	39	40	41	42	43	44	45	46	47	48	49	50	51	52
53	54	55	56	57	58	59	60	61	62	63	64	65	66	67	68
69	70	71	72	73	74	75	76	77	78	1	2	3	4	5	6
79	80	81	82	83	84	85	86	87	88	7	8	9	10	11	12
89	90	91	92	93	94	95	96	97	98	13	14	15	16	17	18

SG 0   SG 1   SG 2

Figure 3.11   H.264 slice groups.

3.6.4 Redundant Slices

The redundant slices (RS) feature sends a secondary representation of a picture region (typically at lower fidelity) that can be used if primary representation is corrupted or lost, thereby improving the robustness of the video transmission. RS allows secondary representations of the MB to be coded in the bitstream.

3.6.5 Flexible Macroblock Ordering

FMO rearranges the coding order of MBs to reduce the probability that a packet loss will affect a large region of the frame. It enables more effective error concealment by ensuring that neighboring MBs will be available for prediction of a missing MB. FMO may subdivide a video frame into 2–8 SGs. The MBs are arranged in different SGs using an macroblock allocation (MBA) map. The MBA map consists of an identification number that specifies to which SG each MB belongs. This ordering exploits the spatial redundancy in frames. For example, the SGs may be designed such that no MB is surrounded by other MBs from the same SG (i.e., error accumulation in a limited region is avoided). In such cases, even if an entire SG is lost, it is possible to construct the MBs using the surrounding MBs from a different SG. If FMO is not activated, each video frame will comprise a single slice with the MBs in raster scan order. The use of FMO is compatible with any type of interframe prediction. FMO may lead to lower coding efficiency caused by disabling prediction across slice boundaries.

Type 0   Type 1

Figure 3.12   FMO types 0 and 1 with two slice groups.

3.6.6   FMO Types

H.264 defines seven FMO types ranging from 0 to 6. They contain a fixed pattern that can be exploited to reduce the overheads needed for reordering the MBs during decoding. Types 0 and 1 are the most popular and examples of these types using two SGs are shown in Figure 3.12. In FMO type 0, each SG contains a series of MBs in raster scan order before another SG starts. Only the length of each SG is transmitted. The number of SGs and the SG identity are coded in the PPS. Using more SGs increases the FMO overhead. FMO type 1 is most suited for transporting videos in high loss transmission environments and can also help improve the privacy of the coded video. In this case, the frame is divided into two SGs with the MBs rearranged in a checkerboard style. Each SG is sent in a different packet. The location of the MBs in the SG is given in equation (3.1). The MBA map is not needed since the formula is known to both encoder and decoder.

$$\text{MB number} \rightarrow \text{SG number } i \rightarrow ((i \bmod w) + ((i/w)n)/2) \bmod n \qquad (3.1)$$

where n = number of SGs (coded in a PPS);

w = frame width in MBs (coded in a SPS);

"$/$" denotes integer division with truncation.

FMO type 2 uses one or more rectangular SGs and a background SG. This separation allows the partition of different regions of interest within a frame. For example, a face can be coded with more bits whereas the background can be coded with lower quality to maintain the total bit rate. Only the top left MB number and the bottom right MB number of each foreground SG are needed for decoding. Types 3, 4, and 5 employ an SG configuration (coded in the PPS) that can change in every frame by making use of periodic patterns. The SG cycle change is included in the header of each slice to keep track of the current position within the cycle of changes. The cycle change is not included in the PPS to reduce overheads and to reflect possible changes in every frame. Type 6 is the most flexible. It allows any MBA map, which must be explicitly coded into the bitstream.

3.6.7 FMO Overhead

FMO overhead is present for all types because they require the creation of slices, which require slice headers. The overheads for FMO type 1 is considerably larger compared to other types (at least two times larger for two SGs and three times larger with four SGs). The first reason for this increased overhead is the reduction of the coding efficiency when more SGs or more slices are used. In H.264, each slice is designed to be decodable without other slices of the frame to minimize error propagation. This implies intraprediction and MV prediction are not allowed across slice boundaries, which degrade coding efficiency. When two SGs are used, only the upper-right and upper-left neighbors can be part of the same slice. Thus, no neighboring MB in the original frame is coded in the same slice. With four SGs, MBs of the same slice are farther from each other than for two SGs. The use of four SGs results in the creation of three extra slices and the coding efficiency degrades.

The efficiency of the entropy coding depends on the elements that have already been coded. When they are correlated to the next elements, entropy coding becomes more efficient. With several slices, the elements of an SG already coded are not used to choose the coding of the elements. Thus, the use of several slices breaks the entropy encoding and reduces the efficiency. The use of B slices reduces temporal correlation, resulting in higher coding efficiency and hence a higher relative FMO overhead. The relative overhead of FMO type becomes smaller as the QP value decreases. This is because the size of the coded video becomes larger whereas the FMO overheads remain fixed. Thus, the use of FMO is more justified for higher quality videos. A summary of the key FMO features is shown in Table 3.4.

3.6.8 Arbitrary Slice Ordering

Arbitrary slice ordering (ASO) is a complementary tool to FMO. ASO allows decoding of slices from an SG in any order. Like FMO, this can prevent entire regions of a frame from being damaged when burst packet losses occur. In addition, ASO reduces the decoding delay when NAL units are not delivered in sequence. ASO and FMO can be used independently. When only ASO is activated, the slices of a unique SG of the frame are not sent in raster scan order.

Table 3.4   FMO Evaluation When Compared with No FMO.

Parameter	Impact
FMO type	Largest overhead for FMO type 1
Small slices	Increases relative overhead
Use of B slices	Increases relative overhead
High QP values	Increases relative overhead
Compression efficiency	Lower
Video quality	Lower for loss-free environment

3.7   TRANSFORM CODING

A low-complexity version of the DCT has been adopted by H.264. The new transform is a scaled integer approximation to the floating-point DCT that was specified with rounding-error tolerances in previous standards. The integer specification eliminates any mismatch problems between the encoder and decoder in the inverse transform. This is because the rounding errors for the integer entries in the transform are known. Thus, in the absence of quantization or transmission errors, perfect reconstruction of the blocks using the exact inverse transform can be achieved. The integer transform employs orthogonal instead of orthonormal basis functions and embeds the normalization in the quantization of each coefficient. The direct or inverse transform can be computed using only simple integer arithmetic and a few 16-bit integer shifts. Since no multiplication is needed, this significantly reduces complexity. The quantization tables are designed to avoid divisions at the encoder. By using an exact exponential scale for the quantization step size as a function of the QP, this ensures that data can be processed in 16-bit arithmetic.

3.7.1   Transform Types

Three types of integer spatial transforms with square block sizes are defined. In general, smaller transforms perform better in areas with discontinuities and block boundaries because they produce fewer ringing artifacts. The $T_{4 \times 4}$ transform, as shown in equation (3.2), is applied to all luma and chroma samples of all intra and inter-predicted error blocks. An 8×8 integer transform is defined for the high profile in equation (3.3). The 8×8 integer transform is more complex than the 4×4 transform but simpler than the 8×8 DCT transform used in H.262. H.264 is built around the 4×4 transform. Each color component of the prediction residual signal is subdivided into smaller 4×4 blocks. While these chroma blocks are generally coded using the 4×4 transform, the transform size can be extended to 8×8. Similarly, the transform for the luma samples can be extended to 16×16:

$$T_{4 \times 4} = \begin{bmatrix} 1 & 1 & 1 & 1 \\ 2 & 1 & -1 & -2 \\ 1 & -1 & -1 & 1 \\ 1 & -2 & 2 & -1 \end{bmatrix} \tag{3.2}$$

$$T_{8 \times 8} = \begin{bmatrix} 8 & 8 & 8 & 8 & 8 & 8 & 8 & 8 \\ 12 & 10 & 6 & 3 & -3 & -6 & -10 & -12 \\ 8 & 4 & -4 & -8 & -8 & -4 & 4 & 8 \\ 10 & -3 & -12 & -6 & 6 & 12 & 3 & -10 \\ 8 & -8 & -8 & 8 & 8 & -8 & -8 & 8 \\ 6 & -12 & 3 & 10 & -10 & -3 & 12 & -6 \\ 4 & -8 & 8 & -4 & -4 & 8 & -8 & 4 \\ 3 & -6 & 10 & -12 & 12 & -10 & 6 & -3 \end{bmatrix} \tag{3.3}$$

3.7.2 Hadamard Transforms

For intrapredicted 16×16 MBs, the $H_{4\times4}$ transform is applied in addition to the $T_{4\times4}$ transform. It transforms all 16 (4×4) DC coefficients of the already transformed blocks containing the luma samples (Figure 3.13). Note that the $T_{4\times4}$ transform closely resembles the orthonormal $H_{4\times4}$ transform, which is a symmetric matrix. The $H_{2\times2}$ transform in equation (3.5) is also a Hadamard transform. It is used to transform the 4 DC coefficients of each 4:2:0 chroma component (Figure 3.13). For 4:4:4 chroma components, $H_{4\times4}$ is used. For 4:2:2 chroma components, both $H_{4\times4}$ and $H_{2\times2}$ are used to perform a 2×4 DC secondary transformation. The DC components for the luma block are transmitted first followed by the AC components. Subsequently, the DC components for the chroma blocks are transmitted followed by their AC components.

$$H_{4\times4} = \begin{bmatrix} 1 & 1 & 1 & 1 \\ 1 & 1 & -1 & -1 \\ 1 & -1 & -1 & 1 \\ 1 & -1 & 1 & -1 \end{bmatrix} \tag{3.4}$$

$$H_{2\times2} = \begin{bmatrix} 1 & 1 \\ 1 & -1 \end{bmatrix} \tag{3.5}$$

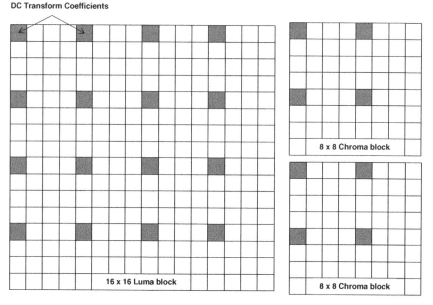

Figure 3.13 DC coefficients for luma and 4:2:0 chroma blocks.

Table 3.5   H.264 Quantization.

QP	0	1	2	3	4	5	6	7	8	9	10	11	12	18	24	30	36	42	48	51
QP Step	0.625	0.6875	0.8125	0.875	1	1.125	1.25	1.375	1.625	1.75	2	2.25	2.5	5	10	20	40	80	160	224

3.7.3   Transform Implementation

Since all transforms in equations (3.2), (3.4), and (3.5) contain only integer numbers ranging from -2 to 2, the forward and inverse transforms can be implemented in 16-bit arithmetic using only a few simple shift, add, and subtract operations. For the Hadamard transforms, only add and subtract operations are necessary. Unlike H.262, which requires a 32-bit (or floating point) transform, a 16-bit transform is sufficient for 8-bit video data using H.264 and is cheap to implement on 16-bit processors (which form the bulk of commonly used digital signal processors). All transform coefficients are quantized by a uniform scalar quantizer with 52 levels. The step size doubles with each QP increment of 6, as shown in Table 3.5. A QP increment of 1 results in an increase of the required data rate of about 12.5%. Similar to prior standards, H.264 also supports weighted quantization matrices.

3.8   ENTROPY CODING

The H.264 entropy coding incorporates context modeling to offer a high degree of statistical adaptivity when coding the video content. The entropy decoder can detect missing MBs in a video frame with no false alarms. Two lossless entropy coding methods are supported in H.264: context-adaptive variable-length coding (CAVLC) and CABAC [3]. Both methods are based on run-level coding. Specifically, the coder first orders the quantized transform coefficients (indexes) of a 2D block in a 1D sequence in decreasing order of expected magnitude. A number of model-based lookup tables are then applied to efficiently encode the nonzero indexes and the length of the tailing zeros. The efficiency of such entropy encoding relies on the fact that nonzero indexes are concentrated at the start of the sequence. Fast, low complexity, and low hardware memory CABAC implementations have been proposed [4].

3.8.1   Context-Adaptive Binary Arithmetic Coding

CABAC employs the statistics of previously coded binary symbols (bins) for estimating the conditional probabilities of current bins. Intersymbol dependencies are exploited by switching between different probability models based on frequency statistics of dominant bins in neighboring blocks. This allows the encoder to adjust to changing symbol statistics for efficient coding. The bins are transmitted using arithmetic coding, which permits a fractional number of bits per symbol. A block diagram of CABAC encoding is shown in Figure 3.14. CABAC incorporates three

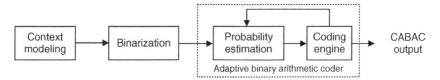

Figure 3.14   Block diagram for CABAC encoding.

stages: context modeling, binarization of syntax elements, and binary arithmetic coding. Context modeling estimates the probability of the bins and maintains different models for each syntax element. For example, MVs and transform coefficients may employ different models. Binarization maps the syntax elements to the bins. In the binarization stage, a binary string is generated for each nonbinary valued syntax element. Finally, arithmetic coding compresses the bins to bits based on the estimated probability, which has been updated during the previous context-modeling stage.

3.8.2   CABAC Performance

CABAC demands more computational power than H.262's entropy encoding, typically about 4:1. Under the VBR mode, CABAC achieves better compression efficiency than CAVLC (typically 5–10%) although CAVLC provides higher decoding throughput. However, the increased decoding time required by CABAC is offset by smaller coded bitstreams due to its superior compression efficiency. Thus, the decoding time for VBR videos using CABAC is comparable to CAVLC for a wide range of video resolutions. With higher resolution or quality VBR videos, the compression gains of CABAC improve over CAVLC. In the CBR mode, the complexity of CABAC becomes evident due to the longer decoding time than CAVLC. This is caused by the enforced CBR rate cap, resulting in no net gains in the compression efficiency and hence the size of the coded bitstreams is more or less similar. The increase in decoding time is insignificant but is more pronounced for higher resolution or quality videos. However, using CABAC for CBR videos leads to better video quality than CAVLC. A summary of the key CABAC performance metrics is depicted in Table 3.6.

3.8.3   Context-Adaptive Variable-Length Coding

CABAC requires a main or high profile and more decoder processing power compared to the other algorithms. For the baseline or extended profiles, the less efficient

Table 3.6   CABAC When Compared with CAVLC.

Mode	Encoding Time	Decoding Time	Compression Efficiency	Video Quality
VBR	Shorter	Slightly longer	Better	Better
CBR	Shorter	Slightly longer	NA	Better

but faster CAVLC must be used. Thus, CAVLC is sometimes used to improve the performance of slower devices. CAVLC employs a single codeword set for all syntax elements except the quantized transform coefficients. The basic approach for coding the transform coefficients is to use the concept of run-level coding as in prior standards. However, the efficiency is improved by switching between VLC tables depending on the values of previously transmitted syntax elements. Unlike CABAC, which is used to code the entire bitstream, only the transform coefficients are coded adaptively in CAVLC. In this case, 32 different VLCs are used. CAVLC is combined with a simpler entropy coding technique called universal variable-length coding (UVLC), which employs exponential Golomb codes to code the syntax elements but not the transform coefficients. Since CAVLC is more efficient than UVLC, it is used to code the quantized transform coefficients. Unlike UVLC, which uses a single VLC table, the VLC tables in CAVLC are designed to match the corresponding data statistics.

3.9   MOTION VECTOR SEARCH

Many coding standards, including H.262 and H.264, treat portions of the video image in blocks, which are often processed in isolation from one another. Regardless of the number of samples in the image, the number of blocks has an impact on the computational requirements. While H.262 employs a fixed block size of 16×16 pixels, H.264 permits simultaneous mixing of seven different block sizes (down to 4×4 pixels). Thus, H.264 can accurately represent fine details with smaller blocks and employ larger blocks for coarse details. For example, patches of blue sky may use larger blocks whereas details of a forest may be coded with smaller blocks.

3.9.1   Motion Search Options

As shown in Table 3.7, H.264 employs variable block sizes in temporal MV search and allows partitions within an MB to be analyzed. This flexibility enables MVs that characterize the displacement of blocks to model the actual motion with high accuracy. Matching blocks can range from the full MB size of 16×16 to 8×8 and 4×4 blocks. Searches may also identify MVs associated with rectangular block sizes of 4×8, 8×4, 8×16, and 16×8. Thus, seven different block sizes can be used

Table 3.7   H.264 Search Types and Block Sizes.

Block Size	Search Type	Impact
8×8	I slice	Enables 8×8 partitions
4×4	I slice	Enables 4×4 partitions
$8 \times 8, 8 \times 16, 16 \times 8$	P slice	Enables 8×8 partitions
$4 \times 4, 4 \times 8, 8 \times 4$	P slice	Enables 4×4 partitions
$8 \times 8, 8 \times 16, 16 \times 8$	B slice	Enables 8×8 partitions

for intermode decision (i.e., determining the best block size for motion estimation). These modes are in addition to the skip mode, direct mode, and two intramodes (INTRA4 and INTRA16) that are also supported in H.264. In contrast, motion estimation in H.262 is more limited since the MVs employ 16×16 MBs. The search options for I, P, and B slices in H.264 are listed as follows.

3.10   MULTIPLE REFERENCE SLICES

Unlike H.262, which restricts the reference slice to the immediate previous slice, H.264 provides additional flexibility for slices to reference up to 16 reference slices. This may be any combination of past and future slices that provide opportunities for more precise interprediction and improved robustness to missing slice data.

3.10.1   Motivations for Using More Reference Slices

The use of more reference slices typically leads to more efficient encoding due to improved interprediction. This may come at the expense of an increase in encoding time. However, using excessive reference slices may not improve coding efficiency significantly since the content of the reference slices that are further away become very different from the slice to be predicted. Using more reference slices can also improve robustness due to corrupted slice data. For instance, partial motion compensation for an interpredicted slice can be applied when one of its reference slices is missing.

3.10.2   Switching Reference Slices

Reference slices can be switched via the switching-predictive (SP) and switching-intra (SI) mechanisms that allow identical reconstruction of slices even when different reference slices or a different number of reference slices are used in the prediction process. These mechanisms avoid the transmission of an I slice and are useful for supporting trick modes, program channel switching, and bitstream decoding other than from the beginning. Since the SI slice is created without using any reference slice, it is more robust but less efficient than the SP slice. The SI slice causes the decoding process to start over, similar to the IDR slice.

3.11   SCALABLE VIDEO CODING

SVC is an H.264 extension that splits a single video bitstream into different representations or layers. This hierarchical layered structure comprises a base layer and two enhancement layers. The media may be scaled down by dropping the enhancement layers. Three scalability types are allowed: temporal, spatial, and video quality. A hierarchy descriptor is used to define the link between the representations. This allows different video representations to be extracted from the bitstream. For example, the lowest video quality can be provided by the base representation. Each additional enhancement representation improves the video quality. Devices with different computational power can select the appropriate representation (i.e., base layer only or base plus enhancement layers) when decoding the video bitstream.

3.11.1 Temporal Scalability

Temporal scalability allows the frame rate of the video bitstream to be varied using interlayer prediction. Because parts of the bitstream are dependent, high-level syntax and interpredicted frames can be reconstructed. This involves generating two video layers. The lower one is encoded by itself to provide the basic temporal rate. The enhancement layer is encoded with temporal prediction with respect to the lower layer. When these layers are decoded and temporally multiplexed, the full temporal resolution of the video source is obtained.

3.11.2 Spatial Scalability

The spatial scalable extension involves generating two spatial resolution video layers from a single video source. The lower layer is encoded by itself to provide the basic spatial resolution. The enhancement layer employs the spatially interpolated lower layer and carries the full spatial resolution of the video source.

3.11.3 Video Quality Scalability

The SNR scalable extension involves generating two video layers of the same spatial resolution but different video quality from a single video source. The lower layer is encoded by itself to provide the basic video quality. The enhancement layer is encoded to enhance the lower layer. When the enhancement layer is added back to the lower layer, it regenerates a higher quality reproduction of the video source.

3.11.4 Disadvantages of SVC

Although SVC provides more flexibility in displaying the video on different devices, it suffers from more overheads than a single-layer bitstream because different layers of the coded video require different levels of error protection. For instance, the base representation and dependent parts of the layered bitstream require more protection than the enhancement representations. In addition, it does not take into account dynamic network conditions when encoding the video. Since all layers (base and enhancement) must be sent to the decoder regardless of its capabilities, transmission bandwidth may not be conserved. The decoder will then decide which layer(s) to display or drop. For HD videos, the base representation already consumes a significant amount of bandwidth, which may lead to visual artifacts and preclude the use of the enhancement representations when bandwidth is low. SVC may be a good fit for cross-layer networking techniques, a rich area of academic research involving PHY/MAC, transport, and application network layers. However, such techniques require many parameters to be optimized and may not be practical for HD video transport. As a result, they have been superseded by HTTP-based adaptive streaming methods.

REFERENCES

1. ITU-T Rec. H.264, *Advanced Video Coding for Generic Audiovisual Services*, January 2012, https://www.itu.int/rec/T-REC-H.264.
2. H.264 Reference Software, http://iphome.hhi.de/suehring/tml.
3. D. Marpe, H. Schwarz, and T. Wiegand, "Context Based Adaptive Binary Arithmetic Coding in the H.264/AVC Video Compression Standard," IEEE Transactions on Circuits and Systems for Video Technology, Vol. 13, No. 7, July 2003, pp. 620–636.
4. H. Malvar, A. Hallapuro, M. Karczewicz, and L. Kerofsky, "Low Complexity Transform and Quantization in H.264/AVC," IEEE Transactions on Circuits and Systems for Video Technology, Vol. 13, No. 7, July 2003, pp. 598–603.

HOMEWORK PROBLEMS

3.1. When compared to SD, HD video requires roughly four times more bandwidth and storage space with H.262. If H.264 is used instead, will this ratio change?

3.2. A H.264 encoder compresses a 7680×4320 video and a UHD-1 video to 90 and 25 Mbit/s respectively. Compute the coding efficiency for both cases, assuming a progressive frame rate of 60 Hz and a sampling format of 4:4:4.

3.3. H.264 uses 52 quantization levels. How many bits are required to represent each level? Compare with the number of levels employed by H.262.

3.4. How is a video with successive black frames coded? List the factors that may affect the video quality for videos coded at the same rate. In general, an encoder must include a decoder to perform prediction and compensation. Will this imply that encoders are always more complex than decoders? Explain why in-loop deblocking filtering is performed at the encoder.

3.5. Table 3.8 shows a comparison of the lossless data coding efficiency using Win-RAR versus lossless H.264 video coding. Even with lossless coding, the H.264 coding efficiency is superior to data compression, especially for HD videos. With judicious selection of coding parameters, more savings can be achieved with lossy coding without sacrificing video quality (Table 3.9). How is it possible that the lossless coding efficiency using H.264 is superior to data compression?

3.6. Some video conferencing vendors claim to achieve a 1000-fold compression efficiency using the baseline profile and a 2000-fold compression efficiency using the main profile. Are these claims valid?

Table 3.8   Efficiency Comparison of Lossless Data and Video Compression.

Video (Progressive Scan)	Coder	Compression Time (min)	File Size (Mbyte)	Efficiency
75s *Dell*, 720 × 480, 30 Hz, 1.18 Gbyte	WinRAR	2.5	443.587	2.6575
uncompressed	H.264	35.5	373.157	3.1591
72s *FCL*, 1280 × 720, 24 Hz,	WinRAR	4.5	706.699	3.3880
2.39 Gbyte uncompressed	H.264	50	387.224	6.1833
125s *Terminator 2*, 1440 × 1080, 25 Hz,	WinRAR	10	1,493.6	4.6356
6.92 Gbyte uncompressed	H.264	126	943.6	7.3376
72s *FCL*, 1920 × 1080, 24 Hz,	WinRAR	10	1,311.0	4.1092
5.39 Gbyte uncompressed	H.264	111	772.1	6.9777
596s *BBB*, 1920 × 1080, 24 Hz,	WinRAR	140	12,288.1	3.6235
44.53 Gbyte uncompressed	H.264	636	4,213.2	10.5682

Table 3.9   Efficiency Comparison of Lossless and Lossy H.264 Coding.

Video (Progressive Scan)	Lossless File Size (Mbyte)	Lossy File Size with Same Video Quality (Mbyte)	Relative Savings
75s *Dell*, 720 × 480, 30 Hz	373.157	19.260	19×
72s *FCL*, 1280 × 720, 24 Hz	387.224	8.637	45×
125s *Terminator 2*, 1440 × 1080, 25 Hz	943.595	32.298	29×
72s *FCL*, 1920 × 1080, 24 Hz	772.062	15.490	50×
596s *BBB*, 1920 × 1080, 24 Hz	4213.150	101.954	41×

3.7. Compare the capabilities of the high and main profiles and explain why the high profile adds minimum complexity to the main profile.

3.8. Video is usually captured in RGB but the color components may be highly correlated. In addition, the human vision is more suited to YC_bC_r. To transform RGB to YC_bC_r, the following equations can be applied with $K_r = 0.2126$ and $K_b = 0.0722$.

$$Y = K_r R + (1 - K_r - K_b)G + K_b B; \; C_b = \frac{1}{2}\left(\frac{B - Y}{1 - K_b}\right); C_r = \frac{1}{2}\left(\frac{R - Y}{1 - K_r}\right)$$

Since the samples are represented using integers, rounding errors may be introduced in the forward and reverse transformations. In addition, the coefficients such as 0.2126 and 0.0722 increases complexity in digital compression and may compromise coding efficiency. Can the first problem be solved by adding two extra bits of precision to the samples? The high profiles in H.264 overcome the second problem by introducing a new color space called YC_gC_o, where C_g and C_o are the green and orange chroma components, respectively.

$$Y = \frac{1}{2}\left(G + \frac{R+B}{2}\right); C_g = \frac{1}{2}\left(G - \frac{R+B}{2}\right); C_o = \frac{R-B}{2}$$

The high profiles overcome the first problem by introducing two sets of equations for the forward and reverse transformations, where $\gg$ denotes an arithmetic right shift operation. Verify that these equations provide exact integer inverses and that an additional bit of sample accuracy is still required.

$$C_o = R - B; \quad C_g = G - B - C_o \gg 1; \quad Y = B + C_o \gg 1 + C_g \gg 1.$$
$$G = Y - C_g \gg 1 + C_g; \quad B = Y - C_g \gg 1 - C_o \gg 1; \quad R = B + C_o.$$

For 4:4:4 video, the high profiles employ residual color transformation to remove the conversion error without significantly increasing the complexity. This requires the use of the RGB domain.

3.9. When coding a video with predictable motion, say a plane flying in straight path, will motion Joint Photographic Experts Group (M-JPEG) be more efficient than H.264? M-JPEG takes a sequence of JPEG coded frames and plays them in order, without interprediction or motion compensation.

3.10. Explain whether there are three slices or three SGs in Figure 3.15.

1	2	3	4	5	6	7	8	9	10	11	12	13	14	15	16
1	2	3	4	17	18	19	20	21	22	23	24	25	26	27	28
5	6	7	8	9	10	11	12	13	14	15	16	17	18	19	20
21	22	23	24	25	26	27	28	29	30	31	32	33	34	35	36
37	38	39	40	41	42	43	44	45	46	47	48	49	50	51	52
1	2	3	4	5	6	7	8	9	10	53	54	55	56	57	58
11	12	13	14	15	16	17	18	19	20	21	22	23	24	25	26
27	28	29	30	31	32	33	34	35	36	37	38	39	40	41	42

Figure 3.15 Slice and SG identification.

3.11. Explain why the video quality improves more with RDO when operated under the CBR mode than the VBR mode. Suppose an RDO algorithm minimizes only the distortion at the expense of a very high coded rate. In other words, only the distortion is optimized. Explain whether such an algorithm could still be useful.

3.12. Will FMO be useful for a channel with bursty losses that corrupt large portions of frames? Is its use justified if the packet loss rate is low, say below 1%? Note that short packets carrying short slices are less prone to errors. Explain why FMO overheads are higher if there are fewer slices in a frame (i.e., many MBs per slice). Note that the coding efficiency improves for this case.

4

H.265/HEVC STANDARD

The emergence of 4K ultra-high definition (UHD) video services has created a stronger need for coding efficiencies superior to H.264/AVC. The ubiquity of online video streaming and HD-capable mobile devices has also increased the demand for efficient and high-quality video delivery. Two key differences between H.264 and H.265 include increased modes for intraprediction and adaptive partitioning for interprediction. Unlike the fixed size macroblock structure of H.264, H.265 defines three different units. The coding unit defines a region in the video frame that shares the same intra or interprediction mode. The prediction and transform units define the regions sharing the same prediction and transform information respectively. The coding and transform units are specified using different quadtrees, which can be flexibly pruned to smaller units to match the characteristics of the video content. The decoding complexity of H.265 is not significantly greater than H.264 and 720p videos can be decoded on H.265-equipped mobile devices without significant battery drain. With H.265, video services are poised to be revolutionized by high quality and scalable delivery of digital content to heterogeneous home and enterprise networks, and user terminals.

4.1 H.265 OVERVIEW

MPEG and Video Coding Experts Group (VCEG) worked together in a partnership known as the Joint Collaborative Team on Video Coding (JCT-VC) to develop the

Next-Generation Video Coding and Streaming, First Edition. Benny Bing.
© 2015 John Wiley & Sons, Inc. Published 2015 by John Wiley & Sons, Inc.

new high efficiency video coding (HEVC) specification. HEVC was subsequently approved as H.265 (ITU-T Recommendation) and MPEG-H Part 2 (ISO/IEC 23008-2). The first version of the H.265 specification, approved in April 2013 [1], lays the foundation by focusing on the 8-bit and 10-bit 4:2:0 color space, which has been widely deployed. Further work in the area of increased bit depth (i.e., 12- and 14-bit encoding) using the 4:2:2 and 4:4:4 color formats was finalized in 2014. H.265 is the latest generation in the MPEG family of video coding standards. The improvement in coding efficiency has been very consistent over the years. Almost every decade, the video coding efficiency has improved by roughly 50%, starting with H.262/MPEG-2 (1994), then H.264/MPEG-4 AVC (2003), and now H.265/HEVC (2013). H.265 allows extensions such as multiview video coding (MVC) and scalable video coding (SVC) to be added.

4.1.1 Fundamental H.265 Benefits

H.265 aims to deliver an average bit rate reduction of 50% for a fixed video quality compared to H.264 or a higher video quality at the same bit rate. Unlike H.264, H.265 supports a flexible coding structure with block sizes that can go up to 64×64 samples. The variable block sizes provide more flexibility in video content adaptation: detailed areas may employ smaller sizes to preserve quality and sharpness whereas coarse regions (e.g., background) can be encoded with larger blocks. The support for larger block sizes may improve compression efficiency significantly, especially when coding high-resolution videos such as UHD videos.

Since high-speed parallel-processing architectures using graphics processing units (GPUs) have become popular in commercial video encoders, H.265 provides improved mechanisms to support parallel encoding and decoding of several frames, including tiles and wavefront parallel processing (WPP). The key enhancements for parallel processing include improved in-loop filters, tiles, and wavefronts (slices). Although GPUs may be expensive, they help simplify hardware implementation because they allow flexible CPU-based software to run on massively parallel hardware and are therefore simple to upgrade. Such a programmable architecture matches well with software defined networks (SDNs) in video transport. More importantly, GPUs significantly improve the speed of video encoding, which can be crucial for supporting real-time TV programming and UHD encoding.

With multithreading, the processing power of modern multicore CPU architectures can be fully utilized even though the H.265 encoder is several times more complex than its predecessors. For example, an increase in processing power in software and hardware is needed to support a doubling of spatial intraprediction sizes, doubling of the number of transforms, almost four times the number of intraprediction directions, flexible quadtree structures, and the increase in the interprediction search space. However, there are some tools that may make H.265 simpler to implement (e.g., entropy coding, deblocking filtering) because the interdependency between some of the processing operations have been removed. Like H.264's use of

Table 4.1   A Comparison of H.264 and H.265.

Function	H.264	H.265
Coding unit	16×16 macroblock	$64 \times 64, 32 \times 32, 16 \times 16$ coding tree unit $64 \times 64, 32 \times 32,$ $16 \times 16, 8 \times 8$ coding unit
Prediction	$16 \times 16, 16 \times 8, 8 \times 16, 8 \times 8,$ $8 \times 4, 4 \times 8, 4 \times 4$	64×64 to 4×4, symmetric/asymmetric
Transform size	$8 \times 8, 4 \times 4$	$32 \times 32, 16 \times 16, 8 \times 8, 4 \times 4$
Transform	DCT	DCT, optional DST for 4×4
Intraprediction	9 modes	35 modes
Luma interpolation	6-tap filter for 1/2 sample followed by bilinear interpolation for 1/4 sample	8-tap filter for 1/2 sample, 7-tap filter for 1/4 sample
Chroma interpolation	Bilinear interpolation	4-tap filter for 1/8 sample
Interprediction	Motion vector	Advanced motion vector prediction (spatial and temporal)
Entropy coding	CABAC, CAVLC	CABAC
In-loop filtering	Deblocking	Deblocking followed by sample-adaptive offset
Parallel processing	Slices, slice groups	Slices, tiles, wavefronts

slices and slice groups, H.265 supports tile-based coding and WPP. Tile-based coding divides a frame into rectangular regions, which can be processed independently. In WPP, a given slice is divided into rows of coding units and each row can be processed separately by a thread, thus providing a finer granularity of parallelism. H.265 decoders are not significantly different from H.264 decoders, even with the addition of an extra in-loop filter. Thus, H.265 software decoding is practical on current hardware.

A comparison of the key features of H.264 and H.265 is shown in Table 4.1. In general, the advantages of H.265 include the following:

- The introduction of larger block structures with flexible subpartitioning mechanisms with block sizes ranging from 64×64 down to 8×8 samples;
- Improved compression efficiency when handling smooth textures in HD and UHD resolutions;
- Improved intra and interprediction accuracy;
- Support for several integer transforms, ranging from 32×32 down to 4×4 samples as well as nonsquare (rectangular) transforms;
- Improved motion information encoding, including a new merge mode, where just an index indicating a previous block is signaled in the bitstream;
- Additional in-loop filtering for improved reconstruction of the decoded signal.

4.1.2 H.265 Applications

H.265 allows service providers to transport higher quality videos with better resolutions at the same bit rates of previous generation codecs, reducing the overall cost of video delivery while improving on the quality of experience for the consumer. The improved quality to bit rate ratio of H.265 enables over-the-top (OTT) providers to quickly jumpstart UHD services with software upgrades. Similarly, it will allow pay-TV providers to expand their online video footprint. However, to deliver UHD or more HD channels over the pay-TV distribution network, hardware decoders and set-top boxes (STBs) may have to be deployed at the customer premise to fully exploit the parallel-processing power of H.265. H.265 may also help improve mobile video streaming if the device is able to decode the videos. Currently, live video calling on smartphones and tablets work well on a limited-range Wi-Fi network with a wireline broadband connection but not on 3G or even 4G networks. Content providers will also enjoy a significant reduction in content library storage, especially video storage for multiscreen and adaptive streaming services. H.265 encoder complexity is manageable with modern multicore CPUs and GPUs. Although more processing power will be needed to decode H.265 videos, the additional decoder power may not be prohibitive for high-end devices (e.g., HDTVs, STBs, PCs) and handheld devices (e.g., tablets, smartphones).

4.1.3 Video Input

Video input to the H.265 encoder is expected to be progressive scan. Interlaced-scanned videos can be deinterlaced prior to encoding. The encoder may indicate that an interlace-scanned video has been sent but H.265 decoders do not support switching between frame and field coding that is required in interlaced video formats. The absence of such tools simplifies the decoder design.

4.2 H.265 SYNTAX AND SEMANTICS

H.265 defines a standard syntax that simplifies implementation and maximizes interoperability. Similar to prior MPEG video coding standards, H.265 only standardizes the bitstream structure, syntax, and constraints such that any H.265 decoder will produce the same output when given a bitstream that conforms to the constraints of the standard. Vendors may optimize specific H.265 feature sets for target applications by balancing compression efficiency, video quality, implementation cost, time to market, and other factors. It employs the general high-level syntax architecture of H.264 that defines frame partitioning, reference frame management, and parameter sets. The syntax also includes support for various types of metadata known as supplemental enhancement information (SEI) and video usability information (VUI). Such metadata provide information about the timing of the video frames, the color space used in the video signal, 3D stereoscopic frame packing information, and display hint information. The VUI can be used to indicate the spatial segmentation of a frame for parallel processing.

4.2.1   Parameter Set Structure

Parameter sets contain information that can be shared for decoding several regions of the video frame. The sequence parameter set (SPS) and picture parameter set (PPS) are both inherited from H.264. These parameter sets have been enhanced by a new video parameter set (VPS) in H.265. The VPS serves as a general higher layer interface for the base H.265 bitstream by providing information about profiles and maximum levels [2]. For the advanced H.265 bitstream, the VPS describes the characteristics of coded video sequences, including the dependencies between temporal sublayers or multiple 3D views. This is more efficient than H.264 multilayer or multiview extensions that employ the SEI message to convey such information.

The video coding layer (VCL) includes all low level signal processing such as block partitioning, intra and interprediction, transform and entropy coding, and loop filtering. The network abstraction layer (NAL) encapsulates coded data and associated signaling information into NAL units, a format that facilitates video transmission systems. The NAL maps the VCL data that represents the content of the video frames onto various transport layers, including RTP/IP, MPEG-4, and MPEG-2 systems and provides a framework for packet loss resilience.

4.2.2   NAL Unit Syntax Structure

Like H.264, a H.265 bitstream consists of a number of access units, each divided into NAL units, including one or more VCL NAL units containing coded video slices and zero or more non-VCL NAL units (e.g., parameter set NAL units or SEI NAL units). Each NAL unit syntax structure is placed into a logical data packet. Each NAL unit includes an NAL unit header and an NAL unit payload. A two-byte NAL unit header is used to identify the purpose of the associated payload data. Information in the NAL unit header can be accessed by media gateways or media aware network elements (MANEs) for stream adaptation. The VCL NAL units identify the slice segments of different frame types for decoder initialization and random access. Table 4.2 shows the NAL unit types.

4.2.3   Reference Frame Sets and Lists

To manage multiple reference frames, a set of previously decoded frames is retained in the decoded picture buffer (DPB) for decoding future frames in the bitstream. Most of the buffer space for H.265 decoding is consumed by the DPB. The size of this buffer is defined by the levels. Another buffer called coded picture buffer (CPB) holds video compressed video data prior to decoding to control the data flow. To help identify the reference frames, the picture order count (POC) is transmitted in each slice header. The set of buffered reference frames is signaled directly using the reference picture set (RPS), which allows generic description of the coding structure. The RPS enables H.265 decoders to detect loss of reference frames.

As in H.264, there are two lists of reference frames (list 0 and list 1) in the DPB. A reference frame index is used to identify a particular frame in one of these lists.

Table 4.2   NAL Unit Types.

Type	Description	Class
0, 1	Slice segment of ordinary trailing frame	VCL
2, 3	Slice segment of TSA frame	VCL
4, 5	Slice segment of STSA frame	VCL
6, 7	Slice segment of RADL frame	VCL
8, 9	Slice segment of RASL frame	VCL
10–15	Reserved for future use	VCL
16–18	Slice segment of BLA frame	VCL
19, 20	Slice segment of IDR frame	VCL
21	Slice segment of CRA frame	VCL
22–31	Reserved for future use	VCL
32	Video parameter set (VPS)	Non-VCL
33	Sequence parameter set (SPS)	Non-VCL
34	Picture parameter set (PPS)	Non-VCL
35	Access unit delimiter	Non-VCL
36	End of sequence	Non-VCL
37	End of bitstream	Non-VCL
38	Filler data	Non-VCL
39, 40	SEI messages	Non-VCL
41–47	Reserved for future use	Non-VCL
48–63	Unspecified (available for system use)	Non-VCL

For uniprediction (i.e., forward prediction), a frame can be selected from either of these lists whereas two frames, one from each list, are selected in biprediction (i.e., forward and backward predictions). When there is only one frame in the list, a value of 0 is assumed for the reference frame index so that it need not be transmitted. If the decoder requires a reference frame list that is different from the initial list, the new list is directly signaled, and not modified from the initial list as in H.264. If a bipredicted block produces the same motion information in list 0 and list 1, the interprediction process in list 1 can be bypassed.

4.2.4   H.265 GOP Structure

Figure 4.1 shows an example of temporal prediction in H.265 GOP. B frames can be used as referenced or unreferenced frames. This is in contrast to H.262 where B frames are unreferenced frames. Figure 4.2 shows the decoding process. The referenced frames (i.e., I, B1, P) must be decoded prior to the unreferenced frames (i.e., b1, b2). In this case, the I frame is used as a reference by more frames (i.e., b1, B1, P) than the referenced P and B frames (e.g., P is a reference for B1, b2, and B1 is a reference for b1, b2). This may help improve interprediction accuracy.

The high-level syntax for identifying the RPS and establishing the reference frame lists for interprediction is more robust to data losses than in H.264, and is more amenable to operations such as random access and trick modes (e.g., fast-forward, rewind, seeking, and adaptive bitstream switching). The key to this improvement is

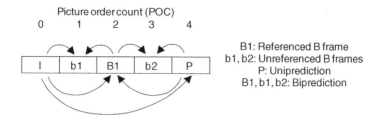

Figure 4.1   Temporal prediction using reference and unreferenced frames.

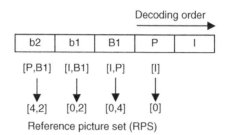

Figure 4.2   Transmitted (and decoding) sequence.

the use of explicit syntax rather than inferences from the stored internal state of the decoding process. Moreover, the associated syntax for these aspects of the design is actually simpler than H.264.

4.2.5   Support for Open GOPs and Random Access

H.265 provides new features to support random access and bitstream splicing, which may be useful for adaptive streaming. Unlike H.264, a conforming H.265 bitstream may or may not start with an instantaneous decoding refresh (IDR) frame. An IDR frame is independently coded and frames that follow it in the bitstream will not reference frames prior to it. In general, these intracoded frames are inserted periodically at the start of a closed GOP. Open GOPs may potentially offer better coding efficiency than closed GOPs. To support open GOPs, H.265 specifies a clean random access (CRA) frame and a broken-link access (BLA) frame, which may be used at any random access point (RAP) in the bitstream. The RAP is a location where a decoder can successfully decode frames without decoding any earlier frames in the bitstream. The CRA frame is a new, independently coded frame that starts at an RAP. In H.265, RAPs are explicitly signaled in the header of an NAL unit, instead of recovery point SEI messages as in H.264. BLA frames originate from bitstream splicing. A CRA frame at the splicing point can be changed to a BLA frame by modifying the NAL unit type.

There is one CRA type and three BLA types. A CRA frame or the first type of BLA frame may be followed by one or two types of leading frames: random access skipped leading (RASL) and random access decodable leading (RADL). RASL frames and RADL frames are leading frames because their display order (i.e., encoder output order) precedes the RAP frame even though they appear after the RAP frame in the decoding order. When random access is performed starting from a CRA frame (which can be at the start or at the middle of a bitstream), the leading frames that appear after the CRA frame may potentially become nondecodable if they require interpredicted reference frames that appear before the CRA frame in the decoding order. Thus, these frames are discarded by the decoder and are designated as RASL frames. Similarly, any RASL frames that are associated with a BLA frame are discarded by the decoder as they may contain references to frames that are not present in the bitstream due to a splicing operation. Like RASL frames, RADL frames may precede a CRA frame in the display order. However, unlike RASL frames, RADL frames do not reference frames that precede the CRA frame in the decoding order. Hence, RADL frames are always decodable. Trailing frames that follow a CRA frame in both decoding order and display order do not contain references to leading frames for interprediction. Hence, trailing frames are always decodable as well. The second type of BLA frames may contain only RADL frames. The third type of BLA frames may not contain RASL or RADL frames. H.265 further specifies two IDR types, which may or may not associate with RADL frames. Thus, there are a total of six RAP frame types.

4.2.6 Video Coding Layer

The H.265 VCL structure is similar to H.264. The basic source-coding algorithm is a hybrid of interframe prediction to exploit temporal statistical dependences, intraframe prediction to exploit spatial statistical dependences, and transform coding of the prediction residual signals to further exploit spatial statistical dependences. The encoder first splits a frame into blocks before intra or interframe prediction is applied. An intracoded frame predicts data spatially from region-to-region within the same frame with no dependence on other frames. An intercoded frame selects a reference frame and an MV to be applied for predicting the samples of each block. The encoder and decoder are able to generate consistent interframe prediction information by using the MV and mode decision data, which are transmitted as side information. The residual of the intraframe or interframe prediction is transformed by a linear 2D spatial transform. The transform coefficients are then scaled, quantized, entropy coded, and transmitted together with the prediction information.

The encoder replicates the decoder processing loop such that both will generate identical predictions for subsequent data. The quantized transform coefficients are constructed by inverse scaling and are then inverse transformed to obtain the approximation of the residual. The residual is then added to the prediction, and the result of that addition may then be fed into one or two loop filters to smooth out blocky artifacts. A copy of the decoded frame is stored in a buffer to be used for the prediction of subsequent frames.

4.2.7 Temporal Sublayers

H.265 employs a hierarchical or scalable temporal prediction structure. The number of decoded temporal sublayers can be adjusted during the decoding process. The location of a temporal sublayer switching point in the bitstream is indicated by temporal sublayer access (TSA) and stepwise temporal sublayer access (STSA) frames. At the location of a TSA frame, it is possible to switch from decoding a lower temporal sublayer to any higher temporal sublayer. Conversely, at the location of an STSA frame, it is possible to switch from decoding a lower temporal sublayer to only one higher temporal sublayer (but not layers above that unless they also contain STSA or TSA frames).

Figure 4.3 shows details of the NAL unit header. The TSA and STSA frames are explicitly signaled using a distinct NAL unit type. The 6 reserved 0 bits carry layer or view type identification as specified by the VPS. This information is meant for the future H.265 decoder that supports multilayer or multiview 3D bitstreams and will be ignored by the base H.265 decoder. H.265 achieves temporal scalability without the need to parse parts of the bitstream using an SEI message. Instead, a temporal identifier (ID) is specified explicitly in the NAL unit header, which indicates the temporal sublayer of the NAL unit. If a slice in a frame with temporal ID of 0 is missing, subsequent frames with temporal ID of 1 and 2 can be skipped until the next RAP. However, if a slice in a frame with temporal ID of 2 is missing, all frames with temporal ID of 0 and 1 can still be correctly decoded. Both temporal layer signaling and stream adaptation (via sub-bitstream extraction) are supported.

4.2.8   Error Resilience

H.264 error resilience features such as slice groups, redundant slices, arbitrary slice ordering, data partitioning, and SP/SI frames, are excluded in H.265. The key H.265 error resilience tool is slicing, which produces VCL NAL units that fit into a data packet, with virtually no coding dependencies to other slices in the same frame. Thus, the loss of a slice may not impact other slices. In addition, the scene information SEI message in H.264 has been retained in H.265 to enable decoders detect scene cuts

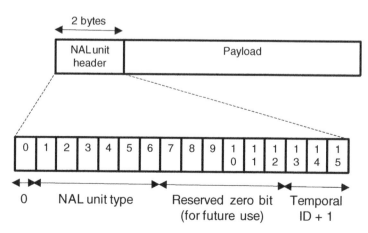

Figure 4.3   NAL unit header.

and gradual scene changes. H.265 further introduces two new SEI messages. The decoded picture hash SEI message contains a checksum derived from the samples of the frame to help detect errors. The structure of pictures (SOP) description SEI message provides information on the interprediction pattern and temporal structure of the bitstream. These messages allow a decoder to determine if the entire frame is lost or if a missing slice causes temporal error propagation. The SOP defines one or more consecutive coded frames in decoding order. The first coded frame is an IDR or CRA reference frame with temporal ID of 0.

4.2.9   RTP Support

The RTP payload format for H.265 reuses many of the concepts of the RTP payload specification for H.264, namely IETF RFC 3984 and its successor RFC 6184 [3]. A single NAL unit can be conveyed in an RTP packet. The NAL unit header may serve as the RTP payload header. Multiple NAL units with identical timestamps (i.e., belonging to the same frame/access unit) can be aggregated into a single RTP packet, and their boundaries are identified using an aggregation payload header. Conversely, an NAL unit can be fragmented into two or more fragments, each of which is transferred in its own RTP packet. In this case, the boundaries of the fragments are identified using a fragmentation header. The RTP payload format supports H.265 parallel processing by indicating the number of cores required for processing the frame. When different parts of the same frame are coded, they may be transmitted using the RTP interleaved mode. The receiver will deinterleave the RTP packets and present the NAL units to the decoder in the correct decoding order.

4.3   PROFILES, LEVELS, AND TIERS

A profile defines a set of coding tools or algorithms used to produce a conforming bitstream. The encoder for a profile may choose which coding tools to use as long as it generates a conforming bitstream. A decoder for a profile must support all coding tools defined in that profile. A level defines limits on certain key bitstream parameters related to the decoder processing load and memory capabilities. These restrictions may include the maximum sample rate, maximum frame size, maximum bit rate, and the minimum compression ratio and capacities of the DPB. Some applications may differ only in terms of maximum bit rate and CPB. To this end, two tiers were specified for level 4 and higher: a Main Tier for general applications and a High Tier for demanding applications. A decoder conforming to a certain tier and level is required to be capable of decoding all bitstreams that conform to the same or lower tier of that level or any level below it.

4.3.1   Profiles

H.265 supports three key profiles namely Main, Main 10, and Main Still Picture. These profiles only support the 4:2:0 sampling format. The Main profile supports a bit depth of 8 bits/sample whereas Main 10 supports 8 and 10 bits/sample. The additional bit depth in Main 10 allows it to provide better video quality than the Main profile. The Main Still Picture profile supports 8 bits/sample and allows only one intracoded video frame using the same constraints as the Main profile. Thus, this profile does not support interprediction.

4.3.2 Levels

H.265 defines 13 levels as shown in Table 4.3. These constraints indicate the required decoder capability to process a bitstream of a specific profile. The luma dimensions may range from 176×144 (quarter common intermediate format, QCIF) to 7680×4320 (8K UHD-2). The maximum resolution for H.262/MPEG-2 is 1920×1152 (incompatible with UHD) and for H.264 is 4096×2304. The H.265 levels are split into two tiers: main (for levels 1–3.1) and high (for levels 4–6.2). The high profiles are designed for demanding applications. A number of levels are similar to H.264. The key difference is the addition of levels 6, 6.1, and 6.2, which define requirements to support 8K UHD-2 videos. The frame length (L) and width (W) are limited according to equation (4.1) where M is the maximum luma frame size:

$$L, W \leq \sqrt{8M} \qquad (4.1)$$

The CPB capacity is computed by multiplying the maximum bit rate by 1s for all levels except level 1, which has a higher CPB capacity of 350,000 bits. The maximum DPB capacity in each level is six maximum-length frames as supported by the level. If the frame size is smaller than the maximum size supported by the level, the DPB

Table 4.3   Main Profile Levels.

Level	Maximum Luma Frame Size (samples)	Maximum Luma Sample Rate (samples/s)	Main Tier Maximum Bit Rate (1000 bit/s)	High Tier Maximum Bit Rate (1000 bit/s)	Minimum Compression Ratio
1	36,864	552,960 (15)	128	–	2
2	122,880	3,686,400 (30)	1,500	–	2
2.1	245,760	7,372,800 (30)	3,000	–	2
3	552,960	16,588,800 (30)	6,000	–	2
3.1	983,040	33,177,600 (33.75)	10,000	–	2
4	2,228,224	66,846,720 (30)	12,000	30,000	4
4.1	2,228,224	133,693,440 (60)	20,000	50,000	4
5	8,912,896	267,386,880 (30)	25,000	100,000	6
5.1	8,912,896	534,773,760 (60)	40,000	160,000	8
5.2	8,912,896	1,069,547,520 (120)	60,000	240,000	8
6	35,651,584	1,069,547,520 (30)	60,000	240,000	8
6.1	35,651,584	2,139,095,040 (60)	120,000	480,000	8
6.2	35,651,584	4,278,190,080 (120)	240,000	800,000	6

Third column parenthesis indicates frame rate in hertz (Hz).

capacity can be increased to up to 16 frames (depending on the chosen frame size). Level-specific constraints are also specified for the maximum number of tiles used horizontally and vertically within each frame and the maximum tile rate (in tiles per second).

Two tiers (main and high) are supported for level 4 and higher (i.e., 8 levels). The highest bit rate of 800 Mbit/s for transporting a single 8K UHD-2 stream using the high tier is over three times greater than the main tier. With a single 802.11ac stream supporting bit rates of up to 866 Mbit/s, such rates can be matched by Wi-Fi networks. However, these rates will be a challenge for online TV providers. Transporting a single 4K UHD-1 stream using the main tier over Wi-Fi or high-speed Internet requires a maximum rate of 25 Mbit/s (30 Hz) and 40 Mbit/s (60 Hz). These rate caps are manageable for servicing online UHD TV.

4.3.3 Range Extensions

The April 2014 draft of the range extensions amendment defines 19 additional profiles: Monochrome 12 (4:0:0), Monochrome 16 (4:0:0), Main 12, Main 4:2:2 10, Main 4:2:2 12, Main 4:4:4, Main 4:4:4 10, Main 4:4:4 12, Monochrome 12 Intra, Monochrome 16 Intra, Main 12 Intra, Main 4:2:2 10 Intra, Main 4:2:2 12 Intra, Main 4:4:4 Intra, Main 4:4:4 10 Intra, Main 4:4:4 12 Intra, Main 4:4:4 16 Intra, Main 4:4:4 Still Picture, and Main 4:4:4 16 Still Picture. Still Picture profiles may use level 8.5, an unbounded level where no limit is imposed on the frame size. Decoders for level 8.5 are not required to decode all level 8.5 bitstreams, since some may exceed their frame size capability.

As shown in Table 4.3, the base value of the minimum compression ratio varies from 2 to 8 depending on the level. This ratio is reduced to half its base value for the 4:2:2 and 4:4:4 sampling formats. Cross-component prediction uses prediction between the chroma/luma components to improve coding efficiency. The reduction in bit rate can be up to 7% for 4:4:4 videos and up to 26% for RGB videos. An RGB video achieves a larger reduction in bit rate due to the greater correlation between the components.

Extended precision processing uses an extended dynamic range for interprediction interpolation and inverse transform. Context adaptive binary arithmetic coding (CABAC) bypass alignment aligns the data before bypass decoding (not supported in range extensions profiles). Intra block copy allows for intraprediction by copying a preceding block region of the frame (not supported in range extensions profiles). The three color planes can be processed independently as three monochrome frames when using 4:4:4 chroma sampling (not supported in range extensions profiles).

4.4 QUADTREES

The adaptive block partitioning structure in H.265 is a key difference compared to previous generations of video coding standards. H.264 uses the 16×16 macroblock (MB) to check for temporal redundancy between successive video frames.

H.265 employs a more flexible quadtree structure [4] that refines motion search. The nested quadtree data structure is based on the basic processing unit called a coding tree unit (CTU) instead of a MB. The CTU or treeblock contains a quadtree syntax that allows for splitting into smaller blocks of variable size depending on the characteristics of the region covered by the CTU. To reconstruct the original image with variable block sizes, the decoder determines the size of each block it receives using the hierarchical quadtree syntax.

As the name implies, a quadtree is a simple tree structure where each node has only four branches. The branch of the quadtree is terminated by a leaf node. At the leaf node, the CTU can be coded in one of three modes: skip, inter, or intra. The decoder may traverse and prune a quadtree easily with marginal increase in decoder complexity. The use of a quadtree enables a frame to be partitioned and coded using variable block sizes depending on the signal characteristics of different regions in the frame. For example, larger blocks can be used for uniform or smooth surfaces and smaller blocks can be used for high detail areas containing high contrast edges. This may lead to more efficient and faster coding of videos, especially if there are large uniform regions in the content.

4.4.1 Variable Block Size Quadtree Partitioning

Each CTU employs its own quadtree that contains several levels. As shown in Figure 4.4, the highest level (the root) represents the largest block and the leaf nodes in the lowest tier represent the smallest blocks. A binary 1 identifies a nonleaf node whereas a binary 0 identifies a leaf node. A quadtree code for each block specifies the CTU size, which is variable. The decision to split a CTU is based on optimizing the rate-distortion performance, which may include minimizing the bit rate and/or maximizing the video quality. For example, if a $2N \times 2N$ CTU contains high correlation between consecutive frames, it should not be split. If a number of rough edges are detected, it should be split into four $N \times N$ blocks. Since a smaller block typically contains lower activity, it can be encoded more efficiently.

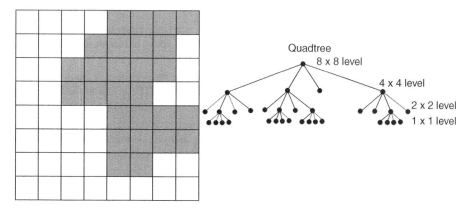

Figure 4.4 Recursive quadtree partitioning.

Figure 4.5 Quadtree block partitioning (blocks expanded for clarity).

In general, large CTUs are recommended for high resolution, low motion, or low quality videos (e.g., videos encoded with high quantization parameter (QP) values). Larger CTUs are especially useful for high-resolution videos since the frames will likely contain smooth regions (Figure 4.5). Large CTUs also reduce overheads due to syntax elements. Quadtree partitioning may lead to redundant sets of motion parameters, which can be avoided by merging the leaves of a specific quadtree using a block merging algorithm. This will enable a single motion parameter set to be generated for a region comprising several colocated motion-compensated blocks.

4.4.2 Coding Tree Units

In H.265 technical nomenclature, a unit is synonymous with a group of pixels whereas a block is synonymous with a group of samples (can be luma or chroma). Thus, three blocks (corresponding to one luma and two chroma components) make up a unit (Figure 4.6). A H.265 frame can be partitioned into square CTUs of up to 64×64 pixels, which is 16 times larger than the 16×16 MB used in legacy video coding standards. A CTU may contain one coding unit (CU) or can be recursively split into smaller CUs. The CU is the basic unit for intra (spatial) or inter (temporal) coding. Each CU is square in shape but can be further partitioned into one or more prediction units (PUs) for intra and interprediction and one or more transform units (TUs) for transform and quantization (Figure 4.7). Unlike CTU partitioning, which may occur multiple times, the splitting of a CU into PUs is done only once. The root of a PU or TU is at the CU level. The decision to code a region of a frame using intra or interprediction is made at the CU level. The shape of the TU depends on the PU. If the PU is square, the TU is also square and its size varies from 4×4 to 32×32 luma

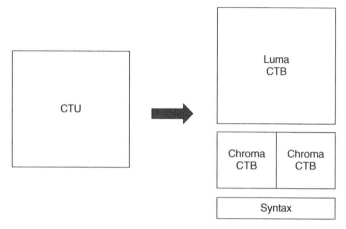

Figure 4.6   Components of a CTU.

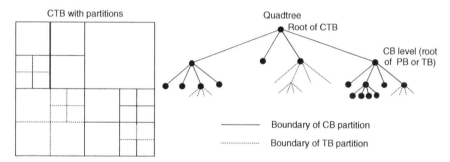

Figure 4.7   Splitting a CTB into CBs and PBs/TBs.

samples. If the PU is nonsquare, the TU is also nonsquare and takes a size of 32×8, 8×32, 16×4, or 4×16 luma samples.

A CTU comprises a luma coding tree block (CTB), the two corresponding chroma CTBs together, and the associated prediction syntax to code the samples as shown in Figure 4.6. Similarly, a CU typically comprises a luma coding block (CB), two chroma CBs, and the prediction syntax. This can be extended to PUs and TUs in the same manner. The tree partitioning in H.265 is applied to both luma and chroma components concurrently. However, exceptions apply when certain minimum sizes are reached for the chroma component.

4.4.3   Splitting of Coding Blocks

The luma and chroma CTBs can be used as CBs or further partitioned into several smaller CBs using tree structures. The appropriate size for the luma CB is determined by the signal characteristics of the region covered by the CTB. The division of a CTU into luma and chroma CBs is signaled jointly using the CTU's quadtree syntax, which specifies the selected size and positions of its luma and chroma CBs. The quadtree splitting process can be iterated until the size for the luma CB reaches the minimum allowable size (8×8 in units of luma samples) that is specified by the encoder via the SPS. The maximum size of a luma CB is the size of the luma CTB. Thus, the size of the CB can range from the size of the CTB (maximum size of 64×64 luma samples) to the minimum size of 8×8 luma samples. The root of the quadtree is at the CTU level. The CU is the decision point to perform intra or interprediction. When the luma CB reaches the minimum size, it is also possible to signal one intraprediction mode for each of its four square subblocks. For the chroma CBs, a single intraprediction mode is selected.

If a 4:2:0 luma CTB comprises $N \times N$ samples, each corresponding chroma CTBs contains $N/2 \times N/2$ samples, where N can be 16, 32, or 64 as specified in the SPS. The larger CTBs may speed up the encoding process for high-resolution videos. More importantly, larger CTBs are key to the compression improvement in H.265, especially if there are no fine details or movement within the CTB. A smaller CTB is comparatively inefficient due to the overheads that will add up since several smaller CTBs (e.g., 16×16) make up a larger CTB (e.g., 64×64). Such dynamic block partitioning within a frame may lead to increased blockiness, which is minimized by H.265 using a two-stage in-loop filter.

4.4.4 Frame Boundary Matching

The frame boundaries are defined in units of the minimum allowed CU (or luma CB) size. This is due to a H.265 mechanism that skips the coding of CUs that fall outside the right and bottom boundaries of the frame. As shown in Figure 4.8, the last few rows of pixels of a 720p frame do not fit complete CTUs of sizes 64×64 or 32×32 because the 720/64 and 720/32 do not give whole numbers. For such cases, CTUs covering regions that lie outside the frame are implicitly split using a CTU quadtree to reduce the CU size until the entire CU fits into the frame. Thus, the frame length and width should be a multiple of minimum CU size (typically 8×8) rather than the CTU size.

4.4.5 Prediction Blocks and Units

A CB may be too large for MVs of small moving objects. In this case, different MVs may be needed depending on the region of the CB. To this end, the luma and chroma CBs can be further split in size and predicted from luma and chroma prediction blocks (PBs). The shape of PUs can be symmetric or asymmetric. Symmetric PUs can be square or rectangular (nonsquare). For example, a CU of size $2N \times 2N$ can be split into two symmetric PUs of size $N \times 2N$ or $2N \times N$ or four PUs of size $N \times N$. Both square and rectangular PUs can be used for interprediction but only square PUs are allowed for intraprediction. Asymmetric PUs are used only for interprediction. As can be seen from Figure 4.9, the shape of a PB need not be square (unlike a CB) and the size may range from 4×4 to 64×64 samples. Interpredicted CUs can apply all

1280

CTU	CTU	CTU	CTU	CTU	CTU	CTU	CTU	CTU	CTU	CTU	CTU	CTU	CTU	CTU	CTU	CTU	CTU	CTU	CTU
CTU	CTU	CTU	CTU	CTU	CTU	CTU	CTU	CTU	CTU	CTU	CTU	CTU	CTU	CTU	CTU	CTU	CTU	CTU	CTU
CTU	CTU	CTU	CTU	CTU	CTU	CTU	CTU	CTU	CTU	CTU	CTU	CTU	CTU	CTU	CTU	CTU	CTU	CTU	CTU
CTU	CTU	CTU	CTU	CTU	CTU	CTU	CTU	CTU	CTU	CTU	CTU	CTU	CTU	CTU	CTU	CTU	CTU	CTU	CTU
CTU	CTU	CTU	CTU	CTU	CTU	CTU	CTU	CTU	CTU	CTU	CTU	CTU	CTU	CTU	CTU	CTU	CTU	CTU	CTU
CTU	CTU	CTU	CTU	CTU	CTU	CTU	CTU	CTU	CTU	CTU	CTU	CTU	CTU	CTU	CTU	CTU	CTU	CTU	CTU
CTU	CTU	CTU	CTU	CTU	CTU	CTU	CTU	CTU	CTU	CTU	CTU	CTU	CTU	CTU	CTU	CTU	CTU	CTU	CTU
CTU	CTU	CTU	CTU	CTU	CTU	CTU	CTU	CTU	CTU	CTU	CTU	CTU	CTU	CTU	CTU	CTU	CTU	CTU	CTU
CTU	CTU	CTU	CTU	CTU	CTU	CTU	CTU	CTU	CTU	CTU	CTU	CTU	CTU	CTU	CTU	CTU	CTU	CTU	CTU
CTU	CTU	CTU	CTU	CTU	CTU	CTU	CTU	CTU	CTU	CTU	CTU	CTU	CTU	CTU	CTU	CTU	CTU	CTU	CTU
CTU	CTU	CTU	CTU	CTU	CTU	CTU	CTU	CTU	CTU	CTU	CTU	CTU	CTU	CTU	CTU	CTU	CTU	CTU	CTU

720

64 × 64 CTUs

Figure 4.8   Fitting CTUs into a 720p frame.

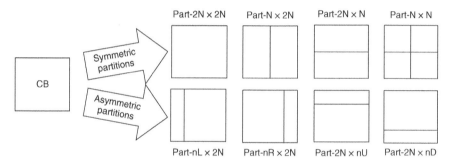

Part-2N × 2N   Part-N × 2N   Part-2N × N   Part-N × N

CB

Symmetric partitions

Asymmetric partitions

Part-nL × 2N   Part-nR × 2N   Part-2N × nU   Part-2N × nD

Figure 4.9   Symmetric and asymmetric PB partitions for interprediction.

modes whereas intrapredicted CUs can only apply the Part-$N \times N$ and Part-$2N \times 2N$ modes.

In the intraprediction mode, the PB size is the same as the CB size (i.e., Part-$2N \times 2N$) for all block sizes except for the smallest CB size. For the latter case, a flag indicates whether the CB is split into four square PBs (i.e., Part-$N \times N$), each with its own intraprediction mode. This enables intraprediction for blocks as small as 4×4 and both luma and chroma components operate with this block size.

In the interprediction mode, the luma and chroma CBs can be split into one, two, or four PBs. The smallest luma PB size is 4×8 or 8×4 samples. The splitting into four PBs is allowed only when the minimum CB size is reached. Suppose the size of a CB is $M \times M$. When a CB is split into four PBs, each PB of size $M/2 \times M/2$ covers a quadrant of the CB. When a CB is split into two PBs, there are six partition possibilities, which can be categorized as symmetric or asymmetric. The size of the two symmetric PBs can be $M \times M/2$ or $M/2 \times M$. The four asymmetric PB partitions employ asymmetric motion partitioning (AMP), which is only allowed when M is 16 or larger for the luma CB. In this case, the size of the PB can be $M/4 \times M$, $3M/4 \times M$, $M \times M/4$, or $M \times 3/4M$. The AMP splitting can also be applied to the chroma CBs in a similar manner. Each interpredicted PB is assigned one or two MVs for uniprediction or biprediction respectively as well as reference frame indices. To reduce memory requirements, 4×4 luma PBs cannot be interpredicted while 4×8 and 8×4 luma PBs are restricted to uniprediction.

4.4.6 Transform Blocks and Units

In the encoding process, an integer DCT is applied on the prediction errors so that these residuals become decorrelated and can be represented by fewer coefficients. The CB may be too large for this purpose because it comprises a detailed (high-frequency) section and a flat (low-frequency) section. Each square CB can be recursively split into four smaller square transform blocks (TBs) using a residual quadtree (RQT). The TBs are the basic units for transform and quantization. Thus, each CB may contain one or more TBs. For an $N \times N$ luma CB, a flag indicates whether it is split into four blocks of size $N/2 \times N/2$. To improve the coding efficiency, the TB need not be aligned with the PB. Hence, unlike previous standards, a single H.265 transform can span across residuals from multiple PBs. Integer transforms are defined for a number of TB sizes. The size and the shape of the TB depend on the size of the PB. The size of square TBs can range from 4×4 to 32×32 luma samples (H.264 supports a transform size of up to 8×8). The size of nonsquare TBs may include 32×8, 8×32, 16×4, or 4×16. The larger transforms typically provide more efficient coding and hardware optimization.

The maximum depth of the RQT in the SPS indicates whether further splitting is possible. Each CB is assigned a flag that indicates whether it is split into four quadrants. The blocks of leaf nodes from the resulting RQT are the TBs. The encoder indicates the maximum and minimum luma TB sizes that it will use. CB splitting is implicit when its size is larger than the maximum TB size. An unsplit CB is implicit when splitting results in a luma TB size that is smaller than the specified minimum. The chroma TB size is half the luma TB size in each dimension, except when the luma TB size is 4×4. In this case, a single 4×4 chroma TB is used for the region covered by four 4×4 luma TBs. For intrapredicted CBs, the decoded samples of the nearest neighboring TBs (within or outside the CB) are used as reference data for intraprediction.

4.4.7 Determining the Quadtree Depth

The flexible block partitioning of H.265 may impose significant computational burden on the encoder when optimizing the various combinations of CU, PU, and TU sizes. Simple tree-pruning algorithms can be used by the encoder to estimate the best partitioning based on rate distortion. For example, the temporal and spatial correlations of the CU depth can be evaluated to eliminate specific depths with minimal impact on rate-distortion performance. Thus, it is unnecessary to do an exhaustive

0	1	4	5
2	3	6	7
8	9	12	13
10	11	14	15

0	1	2	3
4	5	6	7
8	9	10	11
12	13	14	15

Zigzag Scan   Raster Scan

Figure 4.10   Scanning orders for CUs.

search over all possible CU sizes. Flat or homogenous regions in the frame are typically encoded with larger CUs. Areas containing moving objects or object boundaries can be split into small CUs.

4.4.8 Coding Unit Identification

A CU can be identified by an index, which is represented in three different ways: 1D zigzag scan, 1D raster scan, and 2D (x, y) representation. The zigzag and raster scan orders are illustrated in Figure 4.10. Conversion between zigzag scan to 2D (x, y) is quite frequent, especially when locating neighboring CUs. Such conversion is usually performed using table lookups.

4.5 SLICES

A slice can either be an entire frame or a region of a frame (i.e., a frame may be split into one or more slices). A slice is composed of a sequence of CTUs that are decoded in raster scan order (Figure 4.11). Each slice can be transported in a separate NAL unit. Slices are self-contained because prediction data, residual data, and entropy coding are not performed across slice boundaries. Thus, slices within a frame can be parsed and decoded independently from each other. However, when in-loop filtering is applied to a frame, information across slice boundaries may be required but such interslice filtering is optional. A slice breaks the prediction dependency at its boundary, which causes a loss in coding efficiency and may produce visible artifacts at the edges. The different H.265 slice types are listed in Table 4.4. Each block can be coded by one of several coding types depending on the slice type.

Due to the use of variable length codes, synchronization can be a problem when there is data loss in the bitstream. H.264 employs unique a resynchronization marker (0000 0000 0000 0000 1). In H.265, slices are data structures that aid resynchronization and error resilience. The number of CTUs in the slice can be varied depending

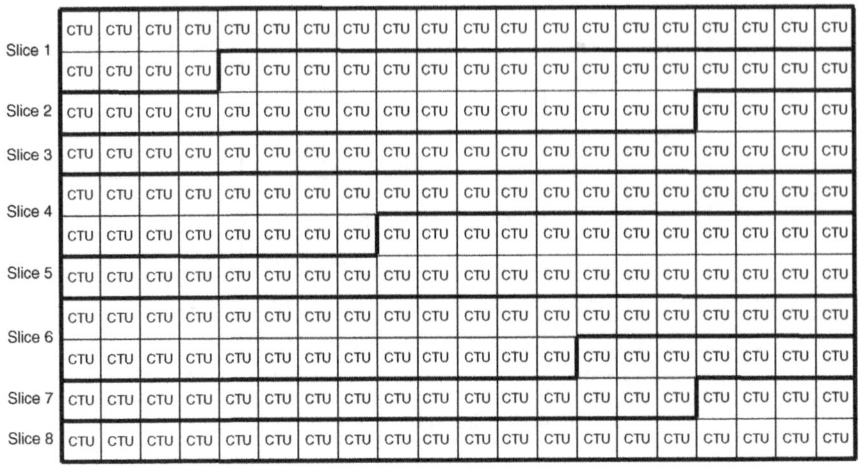

Figure 4.11   An example of slices.

Table 4.4   Slice Types.

Slice Type	Coding Type	Reference Frame List
I	All CTUs are intrapredicted	Not applicable
P	CUs can be intrapredicted or interpredicted with at most one motion-compensated predictor per PB (i.e., uniprediction) and weighted prediction is supported	0
B	CUs can be intrapredicted or interpredicted with at most two motion-compensated predictors per PB (i.e., biprediction)	0 or 1

on the overhead and the activity of the video scene. For I slices, block sizes 32×32, 16×16, 8×8, and 4×4 are supported. For P and B slices, block sizes ranging from 64×64 to 8×4 are supported.

4.5.1   Tiles

Tiles are more flexible than slices in supporting parallel processing. Tiles allow a H.265 frame to be divided into a grid of rectangular regions (each comprising a group of CTUs), which are specified by vertical and horizontal boundaries (Figure 4.12). Each tile must contain at least 256×64 luma samples. The level specifies the limit on the number of tiles. When multiple tiles are encoded in the same slice (using shared header information), they can be independently encoded and decoded simultaneously. Alternatively, a single tile may contain multiple slices. Like slices, tiles break prediction dependences at their boundaries in order to enable independent processing. However, in-loop filtering may cross tile boundaries to control boundary artifacts. Within a frame, consecutive tiles are represented in raster scan order. Similarly, the

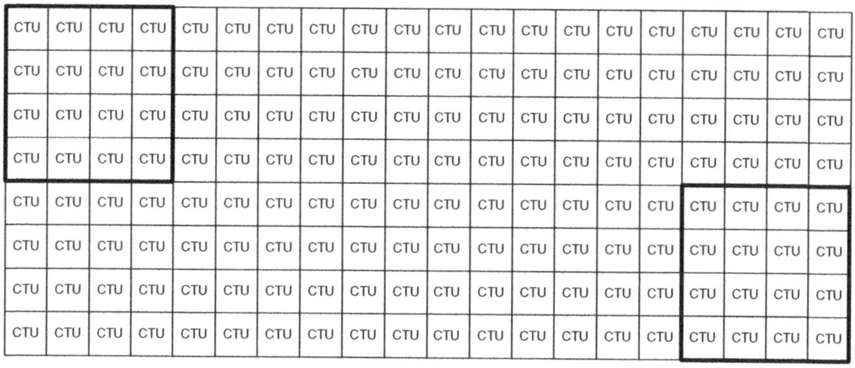

Tile 1 (256 x 256 pixels)

Tile 10 (256 x 256 pixels)

Figure 4.12   An example of square tiles.

CTUs within a tile are processed in raster scan order. When splitting a frame horizontally, tiles may be used to reduce the encoder buffer size since it operates on small sections of the frame. Tiles also allow a frame to employ multiple rectangular sources (e.g., picture-in-picture applications) that are encoded independently.

Tiles enable random access to specific regions of a video frame. Unlike slices, the primary purpose of tiles is to facilitate parallel processing rather than provide error resilience. Tiles address hardware bandwidth restrictions between the codec's processors or cores as well as software-based architectures where the processors or cores are difficult to synchronize on a CTU level without incurring delay or processor under-utilization. Tiles enable parallel-processing architectures for encoding and decoding without the use of complex synchronization threads.

4.5.2   Dependent Slice Segments

Because parallel processing may compromise compression efficiency, dependent slice segments offer an alternative for fast but efficient encoding. Dependent slice segment is a new slice concept specified in H.265 and allows for fragmentation of slices at CTU boundaries. Unlike regular slices (i.e., independent slice segments), dependent slice segments, do not break coding dependencies. Hence, they do not compromise coding efficiency significantly. A dependent slice segment allows specific data related to a wavefront entry point or tile to be accessed more quickly without being encoded as part of the slice. This is achieved by encapsulating a dependent slice as a separate NAL unit. A dependent slice segment for a wavefront entry point can only be decoded after partial decoding of another slice segment has been performed. A dependent slice can also help reduce transmission delay. For example, an encoder can fragment part of a coded video frame into many small units and initiate transmission before coding the rest of the frame without breaking in-frame prediction. H.265 further defines decoding units, which describes parts of the frame as subframes. The H.265 hypothetical reference decoder (HRD) provides a description and the timing of the decoding units.

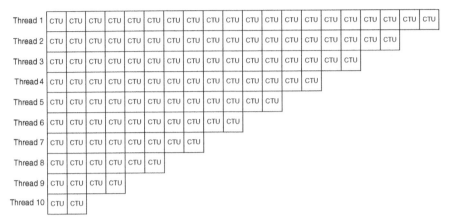

Figure 4.13 Parallel wavefront decoding.

4.5.3 Wavefront Parallel Processing

Slices and tiles both incur a penalty in coding performance since prediction dependencies may cross boundaries and the statistics used in context-adaptive entropy coding must be initialized for every slice or tile. WPP avoids these issues by providing parallel encoding or decoding at the slice level without breaking prediction dependencies and using as much context as possible in entropy encoding. Wavefronts split a frame into rows of CTUs, allowing each row of CTUs to be processed by a different thread as soon as two CTUs have been processed in the row above. Coding dependences between rows are preserved except for the CABAC context state, which is reinitialized at the beginning of each CTU row. Since the intra and interprediction of a CTU may require the top (i.e., directly above the current CTU) and top-right CTU to be available, a shift of two CTUs is enforced between consecutive rows of CTUs that are processed in parallel (Figure 4.13). Hence, unlike tiles, WPP requires additional intercore communication.

The first row of CTUs is processed in the normal way. The second row can start processing after two CTUs have been processed in the first row. Similarly, subsequent rows require the entropy encoder to use data from the preceding row of CTUs. The two-CTU processing delay allows the entropy coder in each row to infer the context models from CTUs in the preceding row. This improves the compression efficiency since the context state is inherited from the second CTU of the previous row, instead of performing a CABAC reinitialization. Thus, each row of CTUs can be assigned to a processing thread in the encoder or decoder and multiple threads can be activated to enable parallel processing.

Because preceding data is used for entropy encoding, WPP may achieve better compression than tiles and avoid visual artifacts that may be induced by using tiles. WPP does not impact single-step processing functions such as entropy coding, predictive coding, and in-loop filtering. For design simplicity, wavefronts are not allowed to operate in conjunction with tiles. However, tiles or wavefronts may employ slices.

Parallel processing using tiles and wavefronts are optional in H.265. Thus, decoders with limited processing capabilities (e.g., smartphones, tablets) may still be able to decode H.265 videos without parallel processing.

4.5.4 Practical Considerations for Parallel Processing

Tiles and WPP allow a parallel encoder or decoder to assign multiple cores or processors to certain spatial parts of the frame. For a decoder processing multiple tiles and wavefronts simultaneously, it must be able to identify the starting position of each tile or slice in the bitstream. This can be achieved with minimum overheads using a table of offsets. In the case of a single-core implementation for tiles, the extra overhead is a result of more complicated boundary condition checking, resetting CABAC for each tile, and performing optional filtering of tile boundaries. There is also the potential for improved data access associated with operating on a subregion of the frame. In a wavefront implementation, additional storage is required to save the CABAC context state between CTU rows and to perform a CABAC reset at the start of each row using this saved state. For multicore implementation, the additional overhead compared to the single-core case relates to memory bandwidth. Since each tile is completely independent, each processing core may decode any tile with little intercore communication or synchronization. A complication is in performing in-loop filtering across the tile boundaries.

A multicore wavefront implementation demands a higher degree of communication between cores and more frequent synchronization operations than a tile-based alternative, due to the sharing of reconstructed samples and mode predictors between CTU rows. The maximum parallel improvement from a wavefront implementation is limited by the ramp-up time required for all cores to become fully utilized and a higher susceptibility to dependency related stalls between CTU rows. All high-level parallelization tools become more effective when encoding or decoding UHD/HD videos. For large frame sizes, it may be useful to enforce a minimum number of frame partitions to guarantee a minimum level of parallelism for the decoder.

4.6 INTRAPREDICTION

H.265 supports various intraprediction modes for all slice types, including intra-angular, intra-planar, and intra-DC. The H.265 intracoding framework is built on spatial sample prediction followed by transform coding and postprocessing. Block-based intraprediction takes advantage of spatial correlation within a frame and attempts to estimate the direction of adjacent blocks with the least error. The improved compression efficiency can be attributed to a quadtree-based variable block size coding structure, block size agnostic angular and planar prediction, adaptive pre- and postfiltering, and transform coefficient scanning based on prediction direction.

Intraprediction operates according to the TB size. Previously encoded boundary samples from neighboring TBs are used as reference for spatial prediction in regions where interprediction is not performed. H.265 defines 35 intraprediction modes, an almost fourfold increase in the number of spatial directions compared to H.264. This increases the number of decision points and the complexity of the codec. The intraprediction mode categories are the same as H.264 namely DC, plane, horizontal, vertical, and directional (angular). The increase is necessary due to the use of larger block sizes in H.265. For example, any one of the 35 intraprediction modes may employ blocks of up to 32×32 samples. The smallest block size remains unchanged at 4×4, which is a computational bottleneck (especially directional prediction) due to the serial nature of intraprediction. In contrast, H.264 employs fewer directions for larger blocks. For example, H.264 defines eight directional intraprediction modes for 4×4 and 8×8 blocks and only four modes for 16×16 MBs.

The 35 modes can be divided into 33 directional prediction modes for square TBs (32×32, 16×16, 8×8, 4×4), planar, and DC. The prediction modes are selected by deriving the most probable directions based on previously encoded adjacent PBs. The chroma prediction mode is either the same as the luma prediction mode (i.e., direct mode) or horizontal, vertical, planar, left-downward diagonal, or DC. For the DC, horizontal, and vertical modes, the simplest modes, an additional postprocess is defined in H.265 where a row and/or column is filtered such as to maintain continuity across block boundaries. The intraprediction mode is applied separately for each TB.

4.6.1 Prediction Block Partitioning

The size of a $2N \times 2N$ intrapredicted CB may be retained by the PB. In this case, the PB is called PART-$2N \times 2N$. Alternatively, the CB is split into four equal-sized PBs called PART-$N \times N$. Since it is also possible to split a CB into four smaller CBs if the size of the current CB is larger than the minimum size, PART-$N \times N$ partitioning is only allowed when the current CB size is the smallest size. Thus, the size of an intrapredicted PB is equal to the size of the CB (i.e., PART-$2N \times 2N$ type) if the CB is not the smallest size. The decision making is shown in Figure 4.14. Although the intraprediction mode is established at the PB level, the actual prediction process operates separately for each TB.

4.6.2 Intra-Angular Prediction

Intra-angular prediction is useful for regions with strong directional edges. As shown in Figure 4.15, H.265 supports a total of 33 prediction directions, which provide more options in choosing the optimal mode that minimizes the prediction error. Because of a large number of angular directions, H.265 intraprediction may provide better coding efficiency than other lossless compression methods such as JPEG, typically by 20–50%. This flexibility increases the search space and the computational latency for intraprediction. The directions are denoted as intra-angular $[k]$, where k is a mode number ranging from 2 to 34.

In order to improve the prediction accuracy and take advantage of the statistically prevalent angles, fine-grained coverage is provided for near-horizontal ($k = 10$) and near-vertical ($k = 26$) angles whereas less dense coverage is provided for near-diagonal angles. In intra-angular prediction, each TB is predicted directionally

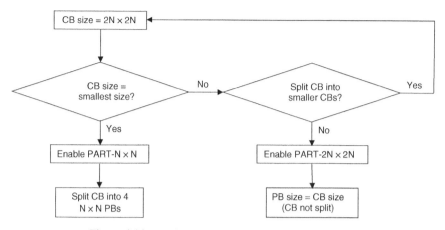

Figure 4.14 Partitioning an intrapredicted CB into PBs.

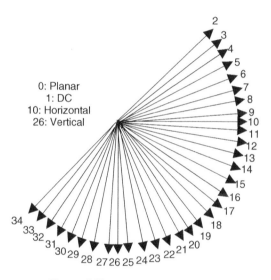

Figure 4.15 Intraprediction modes.

from neighboring samples that are reconstructed but not yet filtered. For an $N \times N$ TB, a total of $4N + 1$ spatial samples may be used for the prediction. The prediction process can involve extrapolating samples from the projected reference sample location according to a given directionality. For intra-angular $[k]$ with $k = 2–17$, the samples located in the above row are projected as additional samples located in the left column. With $k = 18–34$, the samples located at the left column are projected as samples located in the above row. Figure 4.16 shows the intraprediction using angular mode 18.

To improve the intraprediction accuracy, the projected reference sample location is computed with 1/32 sample accuracy. Bilinear interpolation is used to obtain the

Previously encoded
sample from top-left PB

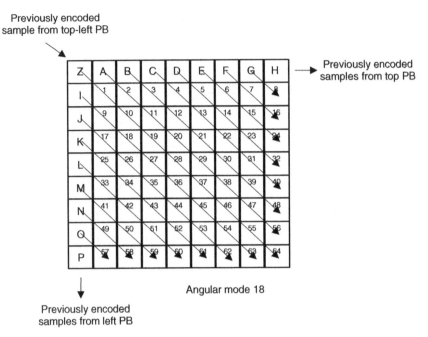

Previously encoded
samples from top PB

Angular mode 18

Previously encoded
samples from left PB

Figure 4.16   Intraprediction using angular mode 18.

value of the projected reference sample using two closest reference samples located at integer locations. The prediction process of the intra-angular modes is consistent across all block sizes and prediction directions, whereas H.264 uses different methods for its supported block sizes of 4×4, 8×8, and 16×16. This design consistency is desirable since H.265 supports more TB sizes and a higher number of prediction directions compared to H.264.

4.6.3   Intra-DC and Intra-Planar Prediction

Intra-DC prediction averages the values of reference samples for the prediction. It is useful for a flat surface with a value matching the mean value of the boundary samples. Intra-planar prediction averages the values of two linear predictions (horizontal and vertical) using four corner reference samples to prevent discontinuities along the block edges. It is useful for predicting smooth sample surfaces because it attempts to match the amplitude of the surface with the horizontal and vertical slopes derived from the boundaries. In H.265, the intra-planar prediction mode is supported for all block sizes whereas H.264 plane prediction is supported only when the luma PB size is 16×16.

4.6.4   Adaptive Smoothing of Reference Samples

Prior to prediction, the reference samples may be smoothed by filtering. The smoothing process can be applied adaptively depending on the prediction mode. The

Table 4.5   Reference Sample Smoothing and Block Size.

Block Size	Reference Sample Smoothing Application
4×4	Not applied
8×8	Diagonal directions $k = 2$, 18, or 34
16×16	All directions except $k = 9$–11 and $k = 25$–27
32×32	All directions except $k = 10$ and $k = 26$

reference samples used for intraprediction can be filtered by a 3-tap [1 2 1]/4 finite impulse response (FIR) smoothing filter in a manner similar to 8×8 intraprediction in H.264. H.265 applies reference sample smoothing by taking into account the directionality, the amount of detected discontinuity, and the block size. In general, larger blocks present a greater need for reference sample smoothing. The application of reference sample smoothing according to the block size is summarized in Table 4.5. As in H.264, the smoothing filter is not applied to 4×4 blocks. For 8×8 blocks, only the diagonal directions (i.e., $k = 2$, 18, or 34) employ reference sample smoothing. For 16×16 blocks, the reference samples are filtered for most of the directions except the near-horizontal and near-vertical directions (i.e., $k = 9$–11 and $k = 25$–27). For 32×32 blocks, all directions except the horizontal (i.e., $k = 10$) and vertical (i.e., $k = 26$) directions use the smoothing filter. When the amount of detected discontinuity exceeds a threshold, bilinear interpolation from three neighboring region samples is applied to smooth the prediction. The intra-planar mode also uses the smoothing filter when the block size is greater than or equal to 8×8. Smoothing is not used (or useful) for the intra-DC case.

4.6.5   Filtering of Prediction Block Boundary Samples

To remove discontinuities along block boundaries for the three modes intra-DC (mode 1) and intra-angular [k] with $k = 10$ or 26 (exactly horizontal or exactly vertical), the boundary samples inside the TB are replaced by filtered values when the TB size is smaller than 32×32. For intra-DC mode, samples in the first row and column in the TB are replaced by the output of a 2-tap [3 1]/4 FIR filter fed by their original and adjacent reference sample values. In horizontal (intra-angular [10]) prediction, the boundary samples of the first column of the TB are modified such that half of the difference between the neighboring reference sample and the top-left reference sample is added. This will smooth the predictor when the vertical direction shows large variation. In vertical (intra-angular [26]) prediction, the same is applied to the first row of samples.

4.6.6   Reference Sample Substitution

The neighboring reference samples are not available at the slice or tile boundaries. In addition, when a loss-resilience feature known as constrained intraprediction is enabled, the neighboring reference samples inside any interpredicted PB are also considered unavailable to avoid error propagation when prior decoded frame data becomes corrupted. While only the intra-DC prediction mode is allowed for such cases in H.264, H.265 allows the use of other intraprediction modes after substituting nonavailable reference sample values with available neighboring reference sample values.

4.6.7 Mode Coding

H.265 supports 33 intra-angular prediction directions, 1 intra-planar mode, and 1 intra-DC mode for luma prediction. Due to the large number of directional modes, H.265 provides three most probable modes (MPMs) when operating in the luma intraprediction mode, instead of one as in H.264. Among the three modes, the first two are initialized by luma intraprediction modes of available intracoded PBs above and to the left. An unavailable prediction mode is considered to be intra-DC. The PB above the luma CTB is always considered to be unavailable to avoid using a line buffer to store neighboring luma prediction modes.

When the first two MPMs are not equal, the third MPM is set equal to {intra-planar, intra-DC, or intra-angular [26] (vertical)} in this order such that it is not a duplicate of one of the first two MPMs. When the first two MPMs are the same, if the first MPM is set to intra-planar or intra-DC, the second and third MPMs are set to {intra-planar, intra-DC, or intra-angular [26] (vertical)} in this order so that they are not duplicates. When the first two MPMs are the same and the first MPM is set to intra-angular, the second and third MPMs are set to two intra-angular modes with the closest k (angle) to the first. The logic is illustrated in Figure 4.17.

If the current luma prediction mode is one of three MPMs, only the MPM index is transmitted to the decoder. If not, the MPMs are excluded and the index of the current

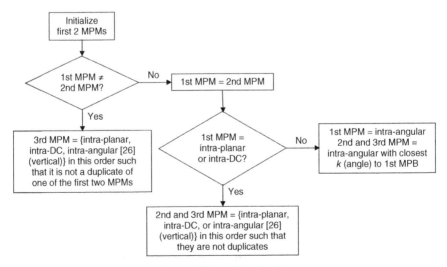

Figure 4.17 MPM selection.

luma prediction mode is transmitted using a 5-bit fixed length code. For chroma intraprediction, the encoder may select one of five modes: intra-planar, intra-angular [26] (vertical), intra-angular [10] (horizontal), intra-DC, and intra-derived. The intra-derived mode specifies a chroma prediction mode with the same angular direction as the luma prediction. Thus, all angular modes specified for luma prediction is also available for chroma prediction. However, the selected chroma prediction mode is coded directly without using MPM prediction.

4.7   INTERPREDICTION

Intercoded frames require reference to other frames. The CBs may perform a final split into smaller PBs, which are interpredicted. Each PB consists of motion parameters such as MV, reference frame index, and reference list flag. H.265 supports more partitions for interprediction than intraprediction. The partitions are listed in Table 4.6 and shown in Figure 4.9. They can be square or nonsquare (rectangular). Nonsquare partitioning offers the possibility of more accurate motion prediction (hence improving compression of motion-compensated residuals) by splitting a CB into two nonsquare PBs. Partitions for intercoded CBs may be symmetric or asymmetric.

An asymmetric motion partition AMP can be applied to CBs of size 64×64 down to 16×16. Unlike symmetric motion partitioning (SMP), AMP is particularly well suited for improving the coding efficiency of irregular object boundaries without further splitting. Nonsquare quadtree transform (NSQT) combines square and nonsquare TBs in a unified transform. It is an extension of RQT (Section 4.8) and supports AMP. However, support for nonsquare shapes requires additional decoder logic as multiple conversions between zigzag scan and raster scan orders may be required.

4.7.1   Fractional Sample Interpolation

Similar to H.264, multiple reference frames may be employed. For each PB, either one or two MVs can be transmitted, giving rise to uniprediction (i.e., single prediction) or biprediction (i.e., combining predictions from two PBs). The PB samples of

Table 4.6   Splitting a CB for Interprediction.

Partition Type	Description
PART-$2N \times 2N$	Not split
PART-$2N \times N$	Split into two equal-size PBs horizontally
PART-$N \times 2N$	Split into two equal-size PBs vertically
PART-$N \times N$	Split into four equal-size PBs (only supported for the smallest allowed CB size)
PART-$2N \times$ nU	Asymmetric split
PART-$2N \times$ nD	Asymmetric split
PART-nL $\times 2N$	Asymmetric split
PART-nR $\times 2N$	Asymmetric split

an interpredicted CB are obtained from a corresponding block region of the reference frame, which is displaced horizontally and vertically by the MV. The MV may have an integer or fractional value. Fractional sample interpolation is used to generate the prediction samples for noninteger sampling locations. Like H.264, H.265 employs MVs with 1/4 sample precision between luma samples in motion compensation. Unlike H.264, the 1/4 samples are calculated from the integer luma samples using a longer filter, instead of using bilinear interpolation based on the neighboring 1/2 and integer samples. For chroma samples, the MV accuracy is determined according to the chroma sampling format. For example, 4:2:0 sampling results in units of 1/8 of the distance between chroma samples. To obtain the noninteger luma samples, separable 1D 8-tap and 7-tap interpolation filters are applied horizontally and vertically respectively to generate the corresponding 1/2 or 1/4 samples (Figure 4.18).

H.265 improves the sample precision using a longer interpolation filter compared to H.264 (7- or 8-tap filtering vs 6-tap filtering). For example, in the interpolation of fractional luma sample locations, H.265 may employ 1/2 sample interpolation with an 8-tap filter or 1/4 sample interpolation with a 7-tap filter. The filters can be applied separately. In contrast, H.264 uses a two-stage interpolation process: 6-tap filtering of 1/2 sample locations followed by linear interpolation of 1/4 sample locations. After the values of one or two neighboring samples at 1/2 sample locations are generated

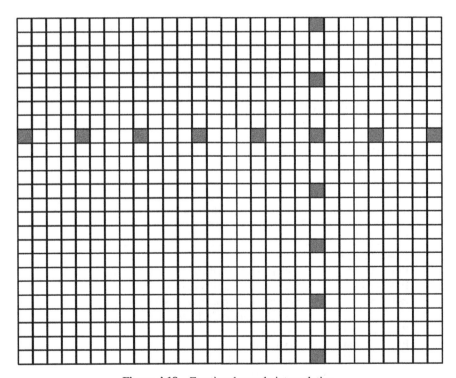

Figure 4.18   Fractional sample interpolation.

using 6-tap filtering, the intermediate results are rounded and the values at integer and 1/2 sample positions are averaged. Thus, luma prediction values at 1/2 sample locations are generated using a 6-tap interpolation filter and prediction values at 1/4 sample locations are obtained by averaging two values at integer and 1/2 sample positions. H.265 uses a single interpolation process for generating all fractional positions without intermediate rounding operations. This improves the accuracy and simplifies the architecture of the fractional sample interpolation. The use of a separable 8-tap filter for luma subsample locations leads to an increase in memory bandwidth and in the number of multiply-accumulate operations required for motion compensation. Filter coefficients are limited to the 7-bit signed range to minimize hardware cost.

Motion compensation in H.265 can be performed using only 16-bit storage elements. As in H.264, weighted prediction using a scaling and offset operation may be applied to the prediction. Whereas H.264 supported both temporally implicit and explicit weighted prediction, only explicit weighted prediction is applied in H.265, by scaling and offsetting the prediction with values sent explicitly by the encoder. The bit depth of the prediction is then adjusted to the original bit depth of the reference samples. The range extensions in H.265 allow high-precision weighted prediction that increases the coding efficiency for fading video scenes at high bit depths.

For uniprediction, the interpolated (and possibly weighted) prediction value is rounded, right-shifted, and clipped to have the original bit depth. For biprediction, the interpolated (and possibly weighted) prediction values from two PBs are added first, and then rounded, right-shifted, and clipped. In H.264, up to three stages of rounding operations are required to obtain each prediction sample for samples located at 1/4 sample locations. If biprediction is used, the total number of rounding operations in the worst case is seven. In H.265, at most two rounding operations are needed to obtain each prediction sample located at 1/4 sample locations. Thus, five rounding operations are sufficient in the worst case when biprediction is used. For an 8-bit depth, the total number of rounding operations in the worst case is further reduced to three. A lower number of rounding operations reduces the accumulated rounding errors.

For fractional chroma sample locations, a set of separable 1D 4-tap DCT-based interpolation filters with the same limitations as the luma filter coefficients can be applied. The fractional accuracy for chroma sample prediction is 1/8 for the 4:2:0 sampling format. The use of a 4-tap filter increases the memory bandwidth and the number of operations compared to H.264 where 2-tap bilinear filtering is used for fractional chroma sample interpolation.

4.7.2 Motion Vector Prediction

In H.264, MVs are encoded using the difference between the reference and predicted MVs. The predicted MV is formed as the median of three neighboring MVs (left, above, and above right). H.264 also has a skip mode, where no motion parameters or quantized residuals are encoded and the motion parameters are inferred from a colocated MB in the previous frame. In H.265, MVs can be predicted either spatially or temporally. For each intercoded PB, the encoder may employ motion parameters, motion merge, or improved skip.

MV information typically consists of horizontal and vertical motion displacement values, and one or two reference frame indices. For prediction regions in B slices, an identification of the reference frame list that is associated with each index is further provided. H.265 defines a signed 16-bit range for both horizontal and vertical MVs. The horizontal and vertical MVs have a range of −32,768 to 32,767 luma samples. With 1/4 sample precision, the MV range becomes −8192 to 8191.75 luma samples. The improvement in MV accuracy is significant compared to H.264, which allows for a horizontal and vertical MV range of −2048 to 2047.75 and −512 to 511.75 luma samples respectively.

4.7.3   Merge Mode

The merge mode allows the MV of a PB to be copied from a neighboring PB (can be temporal or spatial PBs) and no motion parameter is coded. Instead, only the index of the selected candidate in the merge list is coded. The merge mode is similar to the skip and direct motion inference modes in H.264 except for two improvements. H.265 uses index information to select one of several probable MVs as well as information from the reference frame list and reference frame index. In general, the candidate list comprises motion parameters of spatially neighboring blocks and temporally predicted motion parameters that are reused from motion data of a colocated block in a reference frame. In other words, the set of candidates (i.e., previously coded neighboring PBs) are either spatially or temporally close to the current PB. The chosen set of motion parameters is signaled by transmitting an index in the candidate list. Using large block sizes for motion compensation and the merge mode is very efficient for regions with consistent displacements. The temporal candidate can be disabled using a PPS flag. This is useful for wireless applications that are more prone to transmission errors. The B-direct mode of H.264 is replaced by the merge mode since all motion information in the merge mode is derived from the spatial and temporal motion information of the neighboring blocks with residual coding.

The merge candidates consist of spatial neighbor candidates, a temporal candidate, and generated candidates. Figure 4.19 shows the positions of five spatial candidates. The availability of a candidate is checked according to the number on each block as shown in the figure (i.e., block 1 followed by block 2 and so on). The candidate may be omitted for the following reasons:

- The block is intrapredicted or lies outside of the current slice or tile;
- The current PB refers to the first PB within the same CB (the same merge can be achieved by a CB without splitting into prediction partitions);
- The candidate contains the same motion information as another candidate.

For temporal motion, the candidate located in the right bottom position just outside the colocated PB of the reference frame is used if it is available. Otherwise, the

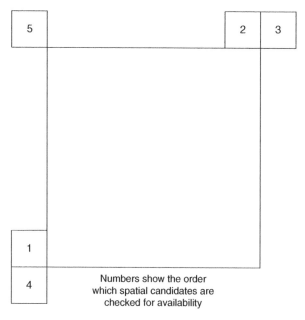

Figure 4.19 Spatial candidate availability.

center position is used. H.265 provides an index to specify the reference frame list that contains the colocated reference frame (and the colocated PU). The amount of the memory to store the motion data of the reference frame is reduced by limiting the dimensions of temporal candidates to a grid of 16×16 luma samples, even if smaller PBs are used at the corresponding location in the reference frame.

The slice header specifies the maximum number of merge candidates. Suppose this number is M and the number of identified merge candidates is N. If $N > M$, only the first $M - 1$ spatial candidates and the temporal candidate are retained. If $N < M$, additional candidates are generated until $N = M$. For a B slice, additional merge candidates are generated by choosing two existing candidates according to a predefined order from reference frame list 0 and 1. For example, the first generated candidate uses the first merge candidate from list 0 and the second merge candidate from list 1. H.265 employs a preconstructed merge candidate list. It specifies 12 candidates in a predefined order but only up to 5 can be selected after removing redundant entries. For a P slice, if $N < M$, zero MVs associated with reference indices from zero to the number of reference frames minus one are used to fill any remaining entries in the merge candidate list.

4.7.4 Skip Mode

In the improved skip mode, the CB is not encoded but directly predicted from the reference frame. Only a skip-flag and index of a motion merge candidate are encoded and transmitted to the decoder. The motion parameters for the current PB are copied from the selected candidate. This allows regions of a frame that change very little between frames or have constant motion (e.g., homogeneous and motionless regions) to be encoded using very few bits. In this case, the skip mode can be considered a special case of the merge mode and all coded block flags are equal to zero. The skip mode is computed first, followed by inter and intra modes. Since the partitioning size of the skip mode is $2N \times 2N$, the PB size is the same as the CB size.

4.7.5 Advanced MV Prediction

If the PB of an interpredicted CB is not coded using the merge or skip modes, advanced MV prediction (AMVP) can be performed. Similar to the merge mode, H.265 allows the encoder to choose the MV predictor among several candidates. The MV is differentially coded using an MV predictor by transmitting an index in the candidate list that specifies the selected predictor and coding a difference vector between the predictor and the actual MV. In this case, data from the reference frame and adjacent PBs is used to determine the most probable MVs. The candidate list is constructed for each MV and the list may include MVs of adjacent blocks with the same reference index as well as a temporally predicted MV.

Only two spatial motion candidates are chosen from five candidates depending on availability. The first candidate is chosen from the locations 4 and 1 in Figure 4.19 and the second is chosen from locations 3, 2, 5 while keeping the search order as indicated. To reduce complexity, H.265 allows a much smaller number of candidates to be used for MV prediction in the nonmerge case. The encoder can always send a coded difference to change the MV. This allows the encoder to focus on motion estimation, which is computationally expensive.

A scaled MV is used when the reference index of the neighboring PB is not the same as the current PB. The neighboring MV is scaled according to the temporal displacements between the current and reference frames as indicated by the reference indices of the neighboring and current PBs. For two spatial candidates with the same MVs, one is discarded. If the number of MV predictors is not equal to two, the temporal MV candidate is included. The temporal candidate is not used when two spatial candidates are available or when temporal MV prediction is disabled. Zero MV(s) can be added until the number of MV predictors is two. A coded flag identifies which MV prediction is used in the nonmerge mode.

4.7.6 Restrictions on Motion Data

H.264 defines restrictions on motion data to reduce memory bandwidth. For example, MVs are downsampled to the 8×8 level and the number of MVs used in consecutive MBs is limited. H.265 defines restrictions that are easier for an encoder to conform to. The smallest luma block sizes are 4×8 and 8×4, and no biprediction is allowed for these blocks. Thus, 4×4 interprediction is prohibited and only the first reference frame list is employed. H.265 further reduces memory requirements by restricting a single MV to a 16×16 MB.

4.7.7 Practical Considerations

For each candidate reference index, the motion search first proceeds over a defined set of integer-sample precision displacement vectors. The measured distortion between the current block and the displaced reference block in the reference frame is indicated by the reference index. For each integer-sample displacement vector, eight surrounding 1/2 sample displacement vectors are evaluated. The 1/2 sample refinement may be followed by a 1/4 sample refinement, where eight 1/4 sample vectors that surround the selected 1/2 sample MV are tested. The distortion measure used for the subsample refinements can be based on the sum of absolute differences (SAD) in the Hadamard domain. The difference between the original block and its motion-compensated predictor is transformed using a blockwise 4×4 or 8×8 Hadamard transform, and the distortion is obtained by summing up the absolute transform coefficients. The use of the SAD in the Hadamard domain usually improves the coding efficiency compared to using the SAD in the sample domain. However, the Hadamard domain should only be used for subsample refinement as it is computationally demanding.

In H.265, the MV predictor for a block is not fixed but selected from a set of candidate predictors. The selected predictor is determined by minimizing the amount of bits to code the MV. The SAD in the Hadamard domain can be used as the distortion measure to select the MV for each reference index. For bipredicted blocks, two MVs and reference indices need to be determined. The initial motion parameters for each reference list are determined independently by minimizing the distortion. This is followed by an iterative step where one MV is held constant and a refinement search is carried out for the other MV. For this iterative refinement, the distortions are calculated based on the predictor that is obtained by biprediction. The decision to code a block using one or two MVs can also be based on the use of SAD in the Hadamard domain as distortion measure. The overall bit rate includes all bits to code the video as well as the motion parameters.

4.8 TRANSFORM, SCALING, AND QUANTIZATION

Block transforms are used to code the intra or interpredicted error residuals. Transform coefficient coding in H.265 covers scanning patterns and coding methods for the last significant coefficient, multilevel significance map, coefficient levels, and sign data. The transforms typically represent integer approximations of a DCT. The integer transforms used in H.265 are better approximations of the DCT than the transforms used in H.264 because the basis vectors of the H.265 transforms have equal energy so there is no need to compensate for the different norms as in H.264. As in H.264, the inverse transforms are specified by exact integer operations. The residual block can be spatially partitioned into square TBs to improve the transform efficiency. H.265 specifies TB sizes of 4×4, 8×8, 16×16, and 32×32. The maximum and minimum TB sizes can be selected by the encoder. The luma CB residual can be represented as a single luma TB or split into four equal-sized luma TBs. If the luma CB is partitioned recursively, each resulting luma TB can be further split into four smaller luma

TBs. Similarly, a chroma CB can be split recursively, as long as it has not reached the minimum size of 4×4. Splitting is implicit when the CB size is larger than the maximum TB size. No splitting is implicit when splitting would result in a luma TB that is smaller than the minimum size. The chroma TB size is half the luma TB size in each dimension, except when the luma TB size is 4×4, in which case a single 4×4 chroma TB is used for the region covered by four 4×4 luma TBs. The adaptive partitioning of a block of residuals to select the transform size is called RQT, with the luma and chroma TBs and associated syntax forming a TU. For each TU, the luma and chroma TBs are each transformed using a 2D integer transform. The splitting depth for each CB is determined by performing RDO over each TU.

For simplicity, only one integer matrix of length 32 points is specified, and subsampled versions are used for other sizes. For example, the matrices for the length-8 and length-4 transforms can be derived by using the first eight entries of rows $0, 2, 4, \ldots$, and the first eight entries of rows $0, 4, 8, \ldots$, respectively of the length-16 transform. Due to the increased size of the transforms, the dynamic range of the intermediate results from the first stage of the transformation should be limited. H.265 explicitly inserts a 7-bit right shift and 16-bit clipping operation after the first 1D inverse transform stage (vertical inverse transform) to ensure that all intermediate values can be stored in 16-bit memory for 8-bit video decoding.

H.264 features 4-point (4×4) and 8-point (8×8) transforms with low implementation complexity using simple sequences of shift and add operations. Since this design strategy does not extend easily to larger transform sizes (e.g., 16-point, 32-point), H.265 defines transforms as fixed-point matrix multiplications (4×4, 8×8, 16×16, 32×32). As the transforms are integer approximations of DCT, they retain the symmetry properties, which facilitate partial butterfly (parallel) implementation. H.265 defines a regular column–row order for the inverse transform. The uniform structure of the matrix multiplication and partial butterfly design make this approach attractive. For example, in software, it is preferable to transform rows, as one entire row of coefficients may easily be held in registers (a row of thirty-two 32-bit accumulators requires eight 128-bit registers, which is implementable on several architectures without register spilling).

Residual differential pulse code modulation (RDPCM) allows vertical or horizontal spatial-predictive coding of residual data in transform skip (TS) and lossless transform bypass blocks (can be selected for use in intra blocks, inter blocks, or both). Transform skip supports block sizes from 4×4 up to 32×32. Transform skip rotation rotates the residual data for 4×4 transform skip blocks. Transform skip context uses a separate context for entropy coding to indicate the blocks that are coded using transform skip.

4.8.1 Alternative 4×4 Transform

For intrapredicted 4×4 luma TB residuals, an integer approximation of the DST can be used as an alternative. The basis functions of DST show a better fit for the weakening statistical correlation (leading to increased residual amplitudes) as the distance from the boundary samples that are used for prediction becomes larger. This is because samples close to the ones used for intraprediction (i.e., near the top or left boundaries) are typically predicted more accurately than the samples further away. Thus, the residuals tend to be larger for samples located further away from the boundaries. The DST is recommended for encoding these kinds of residuals because the basis functions start low and increase. This is in contrast to the DCT basis functions that start high and decrease. The 4-point DST transform is not much more computationally demanding than the 4-point DCT transform but provides only about 1% bit rate reduction in intracoding. The use of the DST transform is restricted to only 4×4 luma transform blocks because for larger block sizes, the additional improvement in coding efficiency is typically marginal.

4.8.2 Scaling

Since the rows of the transform matrix are closely approximated values of uniformly scaled basis functions of the orthonormal DCT, the prescaling operation used in H.264 dequantization is not needed in H.265. The removal of frequency-specific basis function scaling is useful in reducing the intermediate memory size, especially for large transform sizes such as 32×32.

4.8.3 Quantization

H.265 employs uniform reconstruction quantization (URQ). The URQ scheme is controlled by a QP. Like H.264, the QP values range from 0 to 51. If the QP value increases by 6, the quantization step size is doubled. The mapping of the QP values to step sizes is approximately logarithmic. Quantization scaling matrices are supported for various transform block sizes. To reduce the memory needed to store frequency-specific scaling values, only quantization matrices of sizes 4×4 and 8×8 are used. For larger transform sizes of 16×16 and 32×32, an 8×8 scaling matrix is applied by sharing values within 2×2 and 4×4 coefficient groups in frequency subspaces except for DC coefficients, where distinct values are applied. A TB is represented by a flag indicating whether it contains nonzero transform coefficient levels, the location of the last nonzero level in scanning order, a flag for subblocks indicating whether the subblock contains nonzero levels, and syntax elements for representing the chosen set of quantized transform coefficient levels (based on minimum rate distortion).

4.9   ENTROPY ENCODING

After transformation, all syntax elements and quantized transform coefficients are entropy encoded. Unlike H.264, which specifies the context-adaptive variable-length coding (CAVLC) (mandatory) and CABAC (optional) entropy coding methods, CABAC is the only entropy encoding method defined in H.265. The arithmetic coding and context modeling used in CABAC typically achieves better coding efficiency than CAVLC. However, CABAC is more complex. The complexity is

Table 4.7   Syntax Elements.

Syntax Element	Function
skip-flag	Specifies if CB skips interprediction
split-coding-unit-flag	Specifies if CB is further split and coded using context models of spatial neighbors
split-transform-flag	Specifies if TB is further split
cbf-luma	Nonzero transform coefficients for luma component, coded based on splitting depth of transform tree
cbf-cb	Nonzero transform coefficients for C_b component, coded based on splitting depth of transform tree
cbf-cr	Nonzero transform coefficients for C_r component, coded based on splitting depth of transform tree

magnified at higher encoded bit rates (e.g., using small QP values) when transform coefficient data dominates. To improve the CABAC entropy coding time, transform coefficient coding in H.265 is designed for higher throughput than H.264 by reducing data dependencies and feedback loops, thereby facilitating parallel operations. For instance, the number of context-coded binary symbols (bins) has been reduced by eight times and multiple bypass bins can be processed in one cycle. The improved prediction and the larger CB and transform sizes lead to smaller residuals and better coding efficiency. In addition, context modeling and selection in H.265 has been improved to increase efficiency, especially for larger transform sizes. Overall, the improvements include reduced total bins and context-coded bins, grouped bins with the same context and grouped bypass-coded bins, lower context selection dependencies and memory requirements, and support for parallel-processing architectures.

4.9.1   H.265 Binarization Formats

Binarization maps syntax elements to bins. The syntax elements are shown in Table 4.7. Several basic binarization formats are used in H.265, including unary (U), truncated unary, truncated Rice (TR), exponential Golomb (EG), fixed length (FL), and custom. The majority of the syntax elements use these basic binarization formats or some combination for the prefix and suffix. For example, the TR format comprises a combination of truncated unary and FL formats, which is applied by the H.265 encoder to the prefix and suffix of the syntax element respectively. In contrast, H.264 employs the truncated unary prefix followed by an EG suffix. For coefficient values greater than 12, H.265 uses fewer bins in the binarization than H.264. Custom binarization formats (e.g., part mode) are also supported by H.265 syntax elements.

4.9.2   Context Modeling

The goal of context modeling and selection is to accurately model the probability of each bin, which depends on the type of syntax element, the bin index (e.g., most significant bin or least significant bin), and the properties of spatially neighboring coding units. Because several hundred different context models used in H.264 and H.265, a large finite state machine is needed to select the correct context for each bin and update the estimated probability of the selected context after each bin is encoded or decoded. Thus, appropriate selection of context modeling is key to improving the efficiency of CABAC. In H.265, the splitting depth of the coding tree or transform tree is exploited to derive the context model indices of various syntax elements in addition to the spatial neighbors (as used in H.264).

When compared to H.264, H.265 reduces the number of contexts for residual coding by more than half. The reduction in the number of contexts lowers the amount of memory required by the entropy decoder and the cost of initializing the engine. For example, initialization values of the states are defined with 8 bits/context, reduced from 16 in H.264. Although there are substantially fewer number of contexts used in H.265 than in H.264, the entropy coding actually achieves better compression than H.264. In addition, extensive use of the bypass mode of CABAC increases throughput by reducing the amount of data that needs to be coded using CABAC contexts. Another improvement is the modification of the dependencies between the coded data to further maximize throughput.

A popular method for determining contexts in H.264 is to use neighboring spatial samples. For example, the sample value above and to the left of the current block can be used to derive a context for the block. In H.265, such spatial dependencies are mostly avoided to reduce the number of line buffers. In addition, parallel context processing can be enabled where a decoder has the ability to derive multiple context indices in simultaneously. These techniques reduces the entropy decoding bottleneck at high bit rates. In H.264, two interleaved flags are used to signal whether the current coefficient has a nonzero value and whether it is the last one in coding order. This makes it impossible to derive the contexts in parallel. H.265 breaks this dependency by explicitly signaling the horizontal and vertical offset of the last significant coefficient in the current block.

The complexity of entropy decoding using context modeling increases with the video bit rate as more bins need to be processed. To manage this complexity, the bin strings of large syntax elements are divided into a prefix and a suffix. Prefix bins are coded in regular mode (i.e., using context models) whereas suffix bins are coded in a bypass mode. The cost of decoding a bin in bypass mode is lower than in regular mode. In addition, the ratio of bins to bits is fixed at 1:1 for the bypass mode. This ratio is generally higher for the regular mode. In H.264, syntax elements associated with MV differences and quantized transform coefficient levels are binarized using the regular mode. Consequently, level coding becomes a bottleneck at high bit rate as many bits and bins are required. For these syntax elements, the first few bins (i.e., the most significant bins) were context coded while the remaining bins were bypass coded. In H.264, the first nine bins of the MV difference were context coded whereas for H.265, only the first two bins were context coded. Similarly, the number of context-coded bins for each coefficient level was reduced from 14 in H.264 to either 1 or 2 (depending on the number of coefficients per 4×4 block) in H.265.

Overall, for a 16×16 block, the number of context-coded bins in H.265 has been reduced by over eight times compared H.264.

To maximize the use of the bypass mode at high bit rates, H.265 employs a new binarization scheme using Golomb–Rice codes to reduce the worst case number of transform coefficient bins from 15 (in H.264) to 3. When processing large coefficients, the transition boundary between prefix and suffix can be lowered such that a maximum of about 1.5 regular bins need to be processed for each coefficient. This holds for any block of 16×16 transform coefficients.

4.9.3 CABAC Throughput Issues

While CABAC provides high coding efficiency, its bin-to-bin data dependencies is a throughput bottleneck for video encoding. In addition, transform coefficient decoding is the most time consuming part of CABAC. H.265 introduces a new concept of entropy slice that partitions slices into smaller entropy slices which can be processed independently. This allows for parallelizing multiple entropy slices to improve the throughput. However, slice parallelism incurs higher memory resources to buffer all parallel slices, including a huge context table.

4.9.4 CABAC Encoding

The bins are compressed using arithmetic coding. Because multiple bins can be represented by a single bit, this allows syntax elements to be represented by a fractional number of bits, which improves coding efficiency. Arithmetic coding involves recursive subinterval division, where a range is divided into two subintervals based on the probability of the bin that is compressed. The size of each subinterval relative to the initial interval is proportional to the probability of each bin state. The encoded bits represent an offset. When the offset is converted to a binary fraction, it selects one of the two subintervals, which indicates the value of the decoded bin. After every decoded bin, the range is updated to be the same as the selected subinterval, which becomes the new range for the decoding of the next bin, and the interval division process repeats. To efficiently compress the bins to bits, the probability of the bins must be accurately estimated.

Context-coded bins that represent syntax elements are coded using an estimated probability. These probabilities are estimated using context models. They are updated based on the results of previously decoded bins to adapt the estimate to different inputs. Unlike context-coded bins, bypass-coded bins assume an equal probability of 0.5 for both bin states. Hence, bypass bins employ a simpler coding procedure that avoids the feedback loop needed for context selection. In addition, the arithmetic coding becomes faster for bypass-coded bins because the division of the range into subintervals can be done by a shift rather than a lookup table that is required for context-coded bins. Thus, multiple bypass bins can be processed in the same cycle to achieve higher throughput. This can be achieved if the bypass bins appear consecutively in the bitstream. Thus, long runs of bypass bins lead to higher throughput. However, the throughput gains diminish as the number of grouped bins increase.

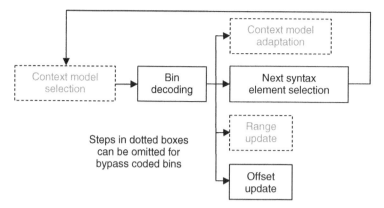

Figure 4.20 CABAC decoding of a context-coded bin.

4.9.5   CABAC Decoding

CABAC decoding is highly sequential due to strict bin-to-bin dependencies (Figure 4.20). The decoding of a bin that is part of a syntax element starts with the context model selection. The context model is used to perform the actual bin decoding. Depending on the result, the range, offset, and context model are updated and the next syntax element is chosen. As the context model selection for the next bin depends on the selection of the corresponding syntax element, it is not possible to overlap the decoding steps for consecutive bins without modifications.

Improving the CABAC decoding speed is important for UHD/HD deployments as well as for handheld personal devices with limited processing power. An effective way to speed up CABAC decoding is to employ a two-stage pipeline [5]. In this case, the selection of the next syntax element is moved to the beginning of the decoding process for the next bin. The first pipeline stage performs the syntax element and context model selection. The data path is duplicated to estimate the possible results of the previously decoded bin. This bin is then forwarded to select the correct context model. The second pipeline stage consists of the bin decoding and range, offset, and context model update.

4.9.6   Coefficient Scanning

Coefficient scanning is performed in 4×4 subblocks for all TB sizes. Thus, only one coefficient region is used for the 4×4 TB size whereas multiple 4×4 coefficient regions are used for larger TBs. Coefficient scanning is dependent on the prediction mode. Three coefficient scanning methods, namely diagonal up-right, horizontal, and vertical scans, are selected implicitly for coding transform coefficients of 4×4 and 8×8 TB sizes in intrapredicted regions. The selection of the coefficient scanning order depends on the intraprediction directions. The vertical scan is used when the direction is close to horizontal. The horizontal scan is used when the direction is close to vertical. For other prediction directions, the diagonal up-right scan is used. The 4×4 diagonal up-right scan is applied only to subblocks of transform coefficients for interpredicted transform coefficients (all TB sizes) and intrapredicted transform coefficients (16×16 and 32×32 TBs).

4.9.7 Coefficient Coding

As in H.264, H.265 sends the location of the last nonzero transform coefficient, a significance map, sign bits, and levels for the transform coefficients. However, there are enhancements to handle the increased TB size. First, the horizontal and vertical frequency locations of the last nonzero coefficient are coded for the TB before sending the significance maps of 4×4 subblocks that identify other transform coefficients with nonzero values. In H.264, a series of last-coefficient identification flags are interleaved with the significance map. The significance map is derived for significance groups connected to the 4×4 subblocks. For all groups with at least one coefficient preceding the last-coefficient location, a significant group flag specifying a nonzero coefficient group is transmitted, followed by coefficient significance flags for each coefficient prior to the indicated location of the last significant coefficient.

The context models for the significant coefficient flags are dependent on the coefficient location as well as the values of the right and bottom significant group flags. Sign data hiding is used to further improve the compression. The sign bits are coded conditionally based on the number and locations of coded coefficients. When sign data hiding is used and there are at least two nonzero coefficients in a 4×4 subblock and the difference between the scan positions of the first and the last nonzero coefficients is greater than 3, the sign bit of the first nonzero coefficient is inferred from the parity of the sum of the coefficient amplitudes. Otherwise, the sign bit is coded normally. At the encoder, this can be implemented by selecting one coefficient with an amplitude close to the boundary of a quantization interval to be forced to use the adjacent quantization interval in cases where the parity would not otherwise indicate the correct sign of the first coefficient. This allows the sign bit to be encoded at a lower cost (in rate-distortion terms) than if it were coded separately – by giving the encoder the freedom to choose which transform coefficient amplitude can be altered with the lowest rate-distortion cost. For each location where the corresponding significant coefficient flag is equal to one, two flags specifying whether the level value is greater than one or two are coded, and then the remaining level value is coded depending on those two values.

4.10 IN-LOOP FILTERS

H.265 emphasizes heavier in-loop filtering than H.264. More specifically, two in-loop filters are applied sequentially to the reconstructed samples before they are stored in the DPB in the decoder loop. The deblocking filter (DBF) is first applied followed by the sample-adaptive offset (SAO) filter. Both in-loop filters are applied within the motion compensation loop (i.e., the filtered image is stored in the DPB as a reference for interprediction). In general, the two-stage filtering in H.265 indirectly helps to improve the video coding efficiency because it allows hierarchical and large block sizes to be used by the encoder within a frame (via adaptive quadtree block partitioning) without introducing visible artifacts at block boundaries. For example, SAO may achieve up to 24% bit rate reduction (about 4% on the average) at the cost of a slight increase in encoding and decoding time (about 1% for encoding and 3% for decoding). SAO is less effective for intracoded frames (about 1% on the average). In addition, video artifacts can be reduced during decoding when parts of a video frame become damaged.

4.10.1 In-Loop Deblocking Filter

The DBF is a low-pass filter adaptively applied at the decoder to block edges to smooth out any blocking artifacts arising from block-based coding. The DBF in H.265 is designed to be simpler than H.264 and to enable support for parallel processing. To this end, DBF in H.265 is performed on a larger block size and is done separately. This reduces the number of computations and enables them to be done in parallel. The DBF is applied to all samples adjacent to a PU or TU boundary except when the boundary is also a frame boundary and when the encoder disables deblocking across slice or tile boundaries. Both PU and TU boundaries are considered because PU boundaries are not always aligned with TU boundaries for some interpredicted CBs. The SPS syntax elements and slice headers determine whether the DBF is applied across slice and tile boundaries.

In H.265, the DBF is only applied to edges on an 8×8 luma or chroma sample grid whereas for H.264, edges lying on a 4×4 sample grid may be filtered. This reduces the number of filter modes by half. A frame may be segmented into 8×8 blocks that can be processed in parallel, as only edges internal to these blocks are filtered. Some of these blocks overlap CTB boundaries and slice boundaries when multiple slices are present. This feature makes it possible to filter slice boundaries in any order without affecting the reconstructed frame.

The processing order of the edges has been modified to enable multiple parallel threads of horizontal or vertical filtering. Horizontal filtering on vertical edges is followed by vertical filtering of horizontal edges. Since vertical edges are filtered before horizontal edges, modified samples from filtered vertical edges can be used to filter horizontal edges. This allows different parallel-processing strategies. For example, all vertical edges can be filtered in parallel, followed by horizontal edges. Alternatively, simultaneous parallel processing of vertical and horizontal edges can be activated, where the horizontal edge filtering process is delayed in a way such that the samples to be filtered have already been processed by the vertical edge filter. It can also be implemented on a CTB-by-CTB basis with only a small processing latency.

H.265 only allows three filter strengths of 0–2 (H.264 uses five). The strength of the DBF is controlled by several syntax elements. Suppose P and Q are two adjacent blocks with a common 8×8 grid boundary. The filter strength of 2 is assigned when one of the blocks is intrapredicted. Otherwise, the filter strength of 1 is assigned if any of the following conditions is satisfied:

- At least one P or Q transform coefficient $\neq 0$;
- Reference index of P $\neq$ reference index of Q;

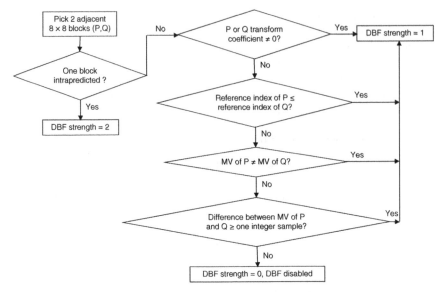

Figure 4.21   Setting the DBF strength.

- MV of $P \neq MV$ of Q;
- Difference between MV of P and $Q \geq$ one integer sample.

If none of the conditions are met, the DBF is disabled using a filter strength of 0. The logic is illustrated in Figure 4.21.

According to the filter strength and the average QP of P and Q, two thresholds t_C and β are obtained from predefined tables. Based on β, luma samples either disable filtering or select weak or strong filtering. This decision is shared across four luma rows or columns using the first and last rows or columns to reduce computational complexity. For chroma samples, no filtering or normal filtering can be enabled. Normal filtering is applied only when the filter strength is greater than one. The addition of clipping in the strong filter mode increases the complexity of the filter.

To summarize, all block boundary types may be filtered in H.265, including CUs, PUs, and TUs (except for 4×4 blocks, which are not filtered to reduce complexity). For each boundary, a decision is made to turn the DBF on or off and whether to apply strong or weak filtering. This decision is based on the sample gradients across the boundary and thresholds derived using the QP in the blocks.

4.10.2   Sample-Adaptive Offset Filter

SAO is a newly added in-loop filter in H.265 to improve picture quality and reduce ringing artifacts. The key idea of SAO is to reduce sample distortion by first classifying reconstructed samples into different categories, obtaining an offset for each category, and then adding the offset to each sample of the category. The classification of each sample is performed at both encoder and decoder to reduce side information. The offset of each category is computed at the encoder and sent to the decoder. To achieve a latency of one CTU, a CTU-based syntax is specified to adapt the SAO parameters for each CTU. A CTU-based optimization algorithm can be used to derive the SAO parameters of each CTU and interleave the parameters into the slice data. Although SAO increases complexity, this is offset by a simplified implementation of the DBF. While the DBF is only applied to samples located at block boundaries, the SAO filter is applied adaptively to all samples satisfying certain conditions (e.g., based on sample gradient).

SAO filtering employs a nonlinear amplitude mapping filter that refines the reconstructed (decoded) samples after the DBF is applied. Unlike the DBF, the SAO filter can enhance the sample reproduction in smooth areas and around edges of a CTB by reducing banding and ringing artifacts. It first classifies the decoded samples based on intensity (i.e., sample amplitude) or edge characteristics. It then minimizes the distortion by adding a band offset (BO) or an edge offset (EO) to each sample for each class of samples. The encoder can also signal that neither BO nor EO is used for a region of a frame. The SAO filter can be disabled or applied in one of two modes as shown in Table 4.8. The offsets are stored in a lookup table that is sent by the encoder. The lookup table can be described by a few parameters that can be determined by statistical histogram analysis at the encoder. The index to the lookup table may be computed according to one of two modes.

BO selects an offset value based on the amplitude of a single sample. The full range of the sample amplitude is divided into 32 equal bands. For 8-bit samples, the bands will be 256/32 or 8 samples wide. Sample values belonging to four of these bands (which are consecutive within the 32 bands) are modified by adding BOs, which can be positive or negative. The offsets are specified for four consecutive bands because in flat (smooth) areas where banding artifacts are common, sample amplitudes in a CTB tend to be clustered in a small range. Because the sample values are quantized to index the table, all samples lying in one band of the value range use the same offset.

EO uses one of four three-sample patterns (directional gradient patterns) to classify samples based on their edge direction. The value of a sample is compared to two of its eight neighbors and classified as one of five categories: minimum (if it is less than two neighbors), maximum (if it is greater than two neighbors), an edge with the sample having the lower value, an edge with the sample having the higher value, or monotonic. For each of the first four categories, an offset is applied. Thus, like BO, EO also uses four offset values. The EO classifications of the CTB are listed in Table 4.9 and illustrated in Figure 4.22. EO requires more computations since the

Table 4.8   SAO Filter Syntax Element sao-type-idx.

Value for sao-type-idx	Function
0	Filter disabled
1	Band offset
2	Edge offset

Table 4.9   SAO Filter Syntax Element sao-eo-class.

Value for sao-eo-class	Direction
0	Horizontal
1	Vertical
2	Diagonal gradient 135°
3	Diagonal gradient 45°

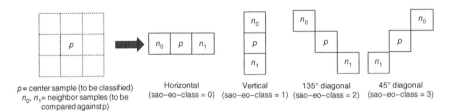

p = center sample (to be classified)
n_0, n_1 = neighbor samples (to be compared against p)

Horizontal (sao-eo-class = 0)   Vertical (sao-eo-class = 1)   135° diagonal (sao-eo-class = 2)   45° diagonal (sao-eo-class = 3)

Figure 4.22   Four directional gradient patterns for edge offset.

Table 4.10   Edge Offset Gradient Patterns.

EdgeIdx	Condition	Meaning
0	Cases not listed below	Monotonic
1	$p < n_0$ and $p < n_1$	Local minimum
2	$p < n_0$ and $p = n_1$ or $p < n_1$ and $p = n_0$	Edge
3	$p > n_0$ and $p = n_1$ or $p > n_1$ and $p = n_0$	Edge
4	$p > n_0$ and $p > n_1$	Local maximum

index is computed based on differences between the current and two neighboring samples.

Each sample in the CTB is classified as one of five EdgeIdx categories by comparing the sample value p at some location with the values n_0 and n_1 of two samples at neighboring locations, as shown in Table 4.10 and Figure 4.23. The EdgeIdx classification is performed for each sample based on decoded sample values, so no additional signaling overhead is required. For EdgeIdx categories 1–4, an offset value from a transmitted lookup table is added to the sample value. The offset values are always positive for categories 1 and 2 and negative for categories 3 and 4. Thus, the filter generally has a smoothing effect.

For SAO types 1 and 2, a total of four amplitude offset values are transmitted to the decoder for each CTB. For type 1, the sign is also encoded. The offset values and related syntax elements (e.g., sao-type-idx, sao-eo-class) are determined by the encoder, typically using metrics related to RDO. The SAO parameters can be inherited from the left or above CTB using a merge flag to enable efficient signaling. Although the operations are simple, SAO may require an additional decoding pass or an increase in line buffers. The offsets need to be derived by an encoder and

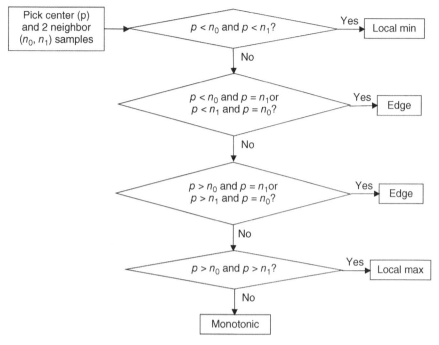

Figure 4.23 Determining an edge offset.

transmitted in the bitstream. The search process in the encoder may require about an order of magnitude more computations than the SAO decoding process.

4.11 SPECIAL H.265 CODING MODES

H.265 defines three special coding modes that can be enabled at the CU or TU level. Table 4.11 summarizes the applications of the three modes:

- Like H.264, IPCM bypasses the prediction, transform coding, quantization, and entropy coding. The samples are directly represented by a predefined number of bits. The purpose is to cater for video characteristics that are unusual (e.g., noise-like signals) and cannot be handled by hybrid coding (i.e., using quadtree block partitioning and motion-compensated prediction).
- Lossless mode sends the intra or interpredicted residual directly into the entropy encoder (using the same neighborhood contexts as the quantized transform coefficients). Thus, transform coding, quantization, and in-loop filtering are bypassed. This reduces the time to encode and decode videos containing complex scenes or full-motion sports. It is also useful for supporting high-quality video applications and services (e.g., tele-surgery). This mode may outperform existing lossless compression solutions such as JPEG 2000 for still images and WinRAR for data archiving. This implies that both still images

Table 4.11 Special H.265 Coding Modes.

Special H.265 Mode	Function	Targeted Videos
I-PCM	Bypasses prediction, transform coding, quantization, entropy coding	Caters for unusual videos (e.g., noise-like signals)
Lossless	Bypasses transform coding, quantization, in-loop filtering	Speeds up encoding/decoding for videos containing complex scenes or full-motion sports, produces high-quality videos
Transform skip	Bypasses transform coding for 4×4 blocks	Improves compression for text, computer graphics, or animated content

and motion pictures can be efficiently compressed by the same encoder. Hence, H.265 can be used for both video editing and video distribution applications.

- Transform skip (TS) only bypasses the transform coding and can only be applied to 4×4 intracoded blocks. The quantization of spatial domain residuals improves the intracoding for video content with sharp details (e.g., text, computer-graphics, animated content).

REFERENCES

1. ITU-T, Recommendation H.265 (04/13), Series H: Audiovisual and Multimedia Systems, Infrastructure of Audiovisual Services – Coding of Moving Video, High Efficiency Video Coding, http://www.itu.int/rec/T-REC-H.265-201304-I.

2. G. Sullivan, J. Ohm, W. Han, and T. Wiegand, "Overview of the High Efficiency Video Coding (HEVC) Standard," IEEE Transactions on Circuits and Systems for Video Technology, Vol. 22, No. 12, December 2012, pp. 1649–1668.

3. T. Schierl, M. Hannuksela, Y. Wang, and S. Wenger, "System Layer Integration of High Efficiency Video Coding," IEEE Transactions on Circuits and Systems for Video Technology, Vol. 22, No. 12, December 2012, pp. 1871–1884.

4. H. Samet, "The Quadtree and Related Hierarchical Data Structures," ACM Computing Surveys, Vol. 16, No. 2, June 1984, pp. 187–260.

5. V. Sze and M. Budagavi, "High Throughput CABAC Entropy Coding in HEVC," IEEE Transactions on Circuits and Systems for Video Technology, Vol. 22, No. 12, December 2012, pp. 1778–1791.

HOMEWORK PROBLEMS

4.1. Is the DBF needed for small block sizes such as 4×4?

4.2. HD display resolutions may lead to incomplete blocks for sizes 16×16, 32×32, and 64×64. Evaluate the minimum number of rows of samples to be added to a 1080p video frame such that there will be no incomplete blocks for these sizes. Will computer display resolutions suffer a similar problem?

4.3. Is the activation of parallelism on a per-frame basis (as opposed to a per-tile basis) useful?

4.4. Show that angular mode 29 requires a PB on the top right for intraprediction but not the PB on the left.

4.5. Complete the decoding order for the GOP structure in Figure 4.24.

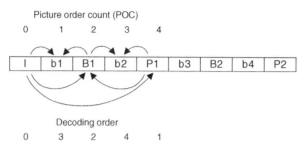

Figure 4.24 GOP structure.

4.6. The sample-based angular intraprediction (SAP) employs the same prediction mode signaling and sample interpolation methods as the H.265 block-based angular prediction but uses adjacent samples for better intraprediction accuracy and performs prediction sample by sample. Explain whether SAP will increase the complexity of H.265 intracoding significantly.

4.7. Motion compensation using multiple reference frames may improve the coding efficiency. Usually, the nearest reconstructed frames are assigned as references. However, the efficiency of such a strategy varies with different content and coding settings. Explore the different ways to assign reference frames and evaluate their efficiencies.

4.8. Which H.265 special mode is most suited for coding sports content with the best efficiency? Is H.265 better suited for OTT service providers than pay-TV providers since OTT services do not carry any sports content?

4.9. Consider the following method of selecting the appropriate PU size based on the motion of the video content. How will this impact the complexity of quadtree pruning?

Mode	Motion Level
Skip	No motion
Inter $2N \times 2N$	Slow motion
Inter $2N \times N$ and $N \times 2N$	Moderate motion
Intra and other inter	High motion

4.10. Evaluate the pros and cons of using a variable size CTU versus a fixed size MB. What content types are more suited for fixed coding blocks?

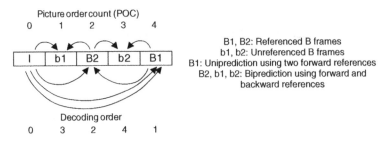

Figure 4.25 Unipredicted B frame using only forward references.

4.11. A unipredicted P frame can be replaced by a referenced B1 frame as shown in Figure 4.25. Such B frames can be predicted using only forward references. For example, a coded block in B1 may employ two MVs and reference indices that point to an I frame. Using two forward references improves the coding efficiency of some video sequences requiring high MV accuracy although this significantly slows down the encoder. Although the B1 frame may be referred as "bipredicted," "unipredicted" is more appropriate since B1 employs only forward prediction and no backward prediction. Hence, the difference between P and B frames becomes less distinct when such unipredicted B frames are employed as reference frames. Forward reference frames are indicated by negative numbers in the RPS. What content types will benefit from unipredicted B frames?

5

ASSESSING AND ENHANCING VIDEO QUALITY

Video quality metrics are used to benchmark or measure the quality of the video as perceived by the user. However, this is not easy since the reconstructed image is not meant to be identical to the original. In low bit rate video coding, perceptual quality assessment assumes a more important role. This chapter covers the basics of video quality assessment and methods to enhance the video quality of videos that are transported over a network.

5.1 INTRODUCTION

In lossy coding, perceptually irrelevant information will be discarded. What counts as "irrelevant" depends on the viewer's subjective response and three measures are normally used: resolution, noise, and overall visual impression. The subjective quality of the reconstructed video, which some people view as the ultimate video quality measure, is not directly taken into account during encoding. Video quality tools typically detect video artifacts caused by codec and packetization errors as well as losses due to network transport but have limited utility in detecting artifacts due to digital capture. The basic video quality metrics can be classified under subjective and objective. Subjective metrics are preferred for assessing the video quality of heavily compressed videos where objective metrics tend to become unreliable. However, subjective gains

Next-Generation Video Coding and Streaming, First Edition. Benny Bing.
© 2015 John Wiley & Sons, Inc. Published 2015 by John Wiley & Sons, Inc.

may not be observable when objective gains are present. Similarly, objective gains may not be observable when subjective gains are present.

5.1.1 Subjective Metrics

Subjective metrics assume a more important role with low-quality videos (e.g., videos coded with high QP values). Subjective metrics assess actual distortions (e.g., blockiness, blurriness, ringing artifacts, added high frequency content, picture outages) as perceived by the viewer. These metrics cannot be measured rigorously using quantitative measures since they take into account the sensitivity and perceptivity of the human visual system (HVS) [1]. Subjective tests are conducted in a controlled environment by collecting the averaged rating or mean opinion score (MOS) from a panel of viewers, which may have different levels of visual clarity. The viewing distance and height of the video display are calibrated against the size of the display. For subjective comparisons, the video sequence under test is presented side-by-side with a sequence generated by the JM or HM reference software. International Telecommunications Union (ITU) recommends longer video sequences (10s where possible) for subjective viewing. ITU also recommends a viewing distance of four to six times the height of the screen [2]. Subjective detection of visual artifacts should be instantaneous.

5.1.2 Limitations of Subjective Metrics

The HVS is complex and may not be consistent across all video displays, resolutions, and human subjects. Children, young adults, and seniors may have varying levels of visual perceptivity. Watching a video up close and from afar may lead to significant differences in subjective video quality. In addition, many subjective tests may not detect subtle artifacts or loss of detail in video playback. When setting up a new HDTV, the HDTV may show two complementary snapshots of the same video, placed side by side to allow the user to select which image view is more desirable. The process is repeated using a few videos. This is a very effective way of selecting the best subjective video quality based on specific user perception. Some video quality measurement methods involve modeling the HVS as accurately as possible. However, human intelligence is still required to prevent false positives. For example, humans can intuitively decipher the age of the movie and an old movie is expected to have poorer quality. Humans can also distinguish between deliberate slow motion (e.g., sports replay), problems in playback, as well as deliberate blurring in background versus blurring or censorship on the subject.

5.1.3 Objective Metrics

Objective metrics are very reliable when assessing the video quality of high-quality videos (e.g., videos coded with low QP values). Objective metrics attempt to quantify the observed video distortions. The quantitative values are typically computed using the original video as a reference and serve as an indicator for possible distortions or artifacts that may be observed in the compressed video. Objective metrics can be further classified as content-independent and content-dependent metrics.

Content-independent metrics do not depend on the type or content of the video and thus, have the same impact on all videos. For example, for a fixed group of pictures (GOP), the number of frames affected by the loss of a P frame remains the same irrespective of the video complexity or resolution. They play an important role in optimizing the network transport of videos. Content-dependent metrics, on the other hand, provide measures whose values depend on the actual video under consideration. Many content-dependent metrics require a reference for their computation. Using the original (perfect) reference video image and the received or compressed (possibly distorted) image, a relative comparison can be evaluated.

5.1.4 Types of Objective Metrics

Tables 5.1 and 5.2 show some content-independent and content-dependent objective metrics respectively. These metrics only identify possible degradation in the video quality and are primarily designed to detect artifacts in still images. Identifying the cause of these problems and providing remedies to fix these problems can be more challenging. For example, the video quality metrics included in ITU-T Recommendation J.247 have outperformed peak signal to noise ratio (PSNR) in validation tests but many typical types of transmission artifacts related to packet losses and the affected frame types are not covered by the tests.

5.1.5 References for Objective Metrics

Many objective video quality approaches are known as full-reference where the similarity of two images are measured using the original distortion-free image as reference. In reduced-reference quality assessment, the reference is only partially available, in the form of a set of extracted features made available as side information

Table 5.1 Content-Independent Metrics.

Time duration	Time duration for which video is affected during a frame loss
Frame size	Size of the frame
Frame type	Type of frame (I, P, or B) that is lost/dropped

Table 5.2 Content-Dependent Metrics.

MSE	Mean squared error in a frame
PSNR*	Peak signal to noise ratio in a frame
Relative PSNR	PSNR as compared to the highest quality video
SSIM*	Structural similarity index

*Full-reference objective metric.

to help evaluate the quality of the distorted image. A third method of video quality assessment does not require any reference. However, such no-reference or "blind" approaches may not be as reliable or accurate compared to the reduced-reference and full-reference metrics. These approaches are also prone to error propagation.

Full-reference metrics are normally computed at the video encoder since it may be impractical to send an undistorted reference video (usually the uncompressed original) to the decoder for comparison. Doing this may require excessive network bandwidth, especially for high resolution videos. On the other hand, no-reference metrics are typically employed by a video decoder at various points in the network at the service provider side. In this case, network bandwidth is only used to transport the compressed video. In general, no-reference video quality metrics cannot perform as well as full-reference video quality metrics.

Many service providers attempt to optimize the video quality and bandwidth consumption by adjusting the coding parameters before sending the coded video across the network. In this case, no visual artifacts are introduced by packet losses or errors attributed to the network and full-reference video quality metrics are sufficient for this purpose. Given that the user will watch the video on the best available screen, the video quality of the same content can be streamlined over different screen sizes.

5.1.6 Network Impact

It can be challenging for a service provider to assess the video quality of videos delivered over its network to its customers because the video content of each frame must be decoded and its quality evaluated. In a point-to-multipoint access network, the usefulness of video quality assessment becomes very limited due to the enormous network overheads needed to track the quality of video playback at thousands of customer premise devices spread across multiple locations. In addition, collecting video quality measurements from these devices and using these measurements to monitor and act on possible distortions require a fair amount of latency. In a one-way broadcast access network (e.g., satellite or digital TV broadcast), video quality measurements become impractical while in mobile access, such measurements may not be updated in a timely manner. Unfortunately, most packet errors and losses occur when the video is transmitted across the access network and this ultimately impacts end-user visual experience. The problem is exacerbated with multicast services when the same video stream is transported under varying link conditions to distributed end-users.

Achieving good video quality requires an understanding of the artifacts generated by coding and network transport. This can be a challenge because the network may not be able to decipher the specific frame within a GOP that is lost and thus take appropriate measures, such as requesting the corrupted frame to be retransmitted. Fortunately, many coding artifacts in legacy codecs have been addressed by newer codecs (e.g., blockiness, frame synchronization, error propagation, color bands) and different frame types can be accorded different priorities by the network. However, digital (source) capture artifacts cannot be detected or corrected by these newer codecs.

5.2 DISTORTION MEASURE

If s_o and s_i are the respective original and reconstructed samples of a block of $M \times N$ samples, the distortion measure D can be defined as

$$D = \sum_{i=1}^{M \times N} |s_o - s_i|^b \qquad (5.1)$$

where $b = 1$ for the sum of absolute differences (SAD) and $b = 2$ for the sum of squared differences (SSD). SSD is a more common distortion metric whereas SAD is used for motion estimation. Hence, all encoders are basically optimized with respect to SSD using the mean squared error (MSE) as defined in (5.2):

$$\text{MSE} = \frac{\text{SSD}}{M \times N} \qquad (5.2)$$

5.2.1 Sum of Absolute Differences

Each block can be compared to the corresponding block in the previous video frame using a similarity metric called SAD as shown in (5.3):

$$\text{SAD}(n) = \sum_{i=0}^{N-1} \sum_{j=0}^{M-1} |F_n(i,j) - F_{n-1}(i,j)| \qquad (5.3)$$

where F_n is the nth video frame of size $N \times M$, and i and j denote the sample coordinates. The encoder can use the SAD values stored in a predefined number of previous frames as a reference to make the best decision. SAD can also be used in temporal error concealment (EC) to identify the best replacement block. It generally ensures spatial continuity and produces visually good results.

5.2.2 Sum of Absolute Transformed Differences

The sum of absolute transformed differences (SATD) is another video quality metric used for block-matching in motion estimation. It employs a frequency transform (e.g., Hadamard transform, which is a generalized version of the Fourier transform) of the differences between the samples in the original block and the corresponding samples in the coded block. A small subblock (e.g., 4×4 samples) is typically used for transformation rather than the full size block (e.g., 16×16 MB). Although SATD is slower than SAD due to the increased complexity, SATD may predict the objective and subjective video quality more accurately.

5.3   PEAK SIGNAL TO NOISE RATIO

The PSNR compares the maximum sample value (the desired signal) with the difference in the sample values of the original uncompressed and decompressed videos (the noise signal). Although there are three components in the PSNR computation (Y, U, V), visual perception is most sensitive to brightness. Hence, the luminance component of PSNR (i.e., Y-PSNR) is the more popular metric. A high PSNR indicates a less noisy signal. PSNR is governed by the MSE and number of bits per sample of the video signal to be coded (B). For two $M \times N$ monochrome images I and K, the PSNR is given by (5.4). For RGB color images, the same equation is valid but the MSE is sum over all squared value differences divided by the image size and by three. Typical PSNR values range from 30 to 50 dB, where a higher value indicates better video quality. Decoded video frames with PSNR values greater than 40 dB normally results in transparent videos (i.e., the decoded frames become indistinguishable from the original frames). The PSNR is infinite when two images are identical since the MSE will be zero. A PSNR of zero corresponds to the case when I is completely white and K is completely black (or vice versa). In general, an image or video frame with significant details or high scene complexity lowers the PSNR since it is more difficult for a lossy encoder to replicate the original frame:

$$PSNR = 20\log_{10}\left(\frac{2^B - 1}{\sqrt{MSE}}\right)$$

$$\text{where MSE} = \frac{1}{M \times N}\sum_{i=0}^{M-1}\sum_{j=0}^{N-1}|I(i,j) - K(i,j)|^2 \tag{5.4}$$

5.3.1   Combined PSNR

In some situations, the rate-distortion curves of the luma and chroma components are combined. The combined PSNR or YUV-PSNR is computed by weighting the PSNR of the sum of the individual components for each frame according to (5.5). Clearly, a greater emphasis is placed on Y-PSNR. Since the chroma components (i.e., U-PSNR, V-PSNR) are typically subsampled (e.g., using the 4:2:0 color format), fewer samples are available for computation when compared to the luma component (i.e., Y-PSNR). Thus, if the QP value is held constant, the PSNR values for the chroma components are inherently higher than the luma component. As demonstrated in Figures 5.1 and 5.2, for complex content such as *Fountain*, the difference in PSNR can be an order of magnitude higher for higher QP values. As expected, there is a high degree of correlation between YUV-PSNR and Y-PSNR in Figures 5.3 and 5.4. Hence, either metric can be used to assess the video quality objectively without introducing significant errors. Y-PSNR is preferred since it is easier and faster to compute than YUV-PSNR:

$$\text{YUV-PSNR} = \frac{6(Y\text{-PSNR}) + (U\text{-PSNR}) + (V\text{-PSNR})}{8} \tag{5.5}$$

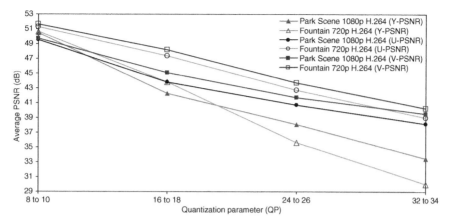

Figure 5.1   Average *Y*, *U*, and *V*-PSNR of decoded H.264 videos.

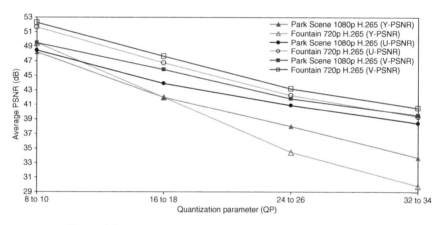

Figure 5.2   Average *Y*, *U*, and *V*-PSNR of decoded H.265 videos.

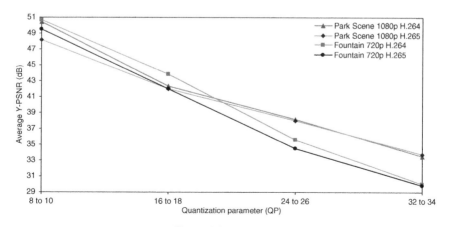

Figure 5.3   *Y*-PSNR.

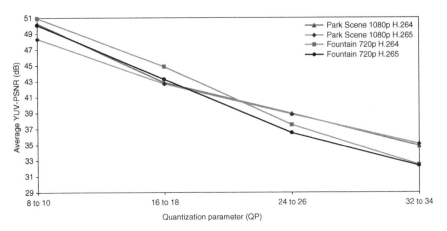

Figure 5.4   YUV-PSNR.

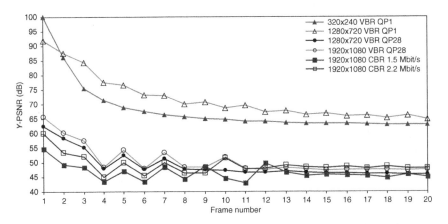

Figure 5.5   *Y*-PSNR for same video with different resolutions and coding settings.

5.3.2   Impact of Video Resolution and QP on PSNR

Figure 5.5 shows the *Y*-PSNR for the same VBR video with different resolutions and coding parameter settings. The first frame is an I frame and all other frames are P frames. The H.264 encoder is used. The same video content with different resolutions may lead to similar *Y*-PSNR if the videos are coded with the same QP value. This is because the video frames for these videos are coded in essentially the same way. For example, the two curves representing the 320×240 and 1280×720 resolutions with a QP value of 1 match up quite closely. A similar trend is observed for the 1280×720 and 1920×1080 resolutions with a QP value of 28. The *Y*-PSNR values are not identical because a larger number of coding blocks, transform blocks, and MVs are needed to code a higher-resolution video. On the other hand, if the same video content with different resolutions is encoded using different QP values, the

Y-PSNR values will be different. For low QP values, the Y-PSNR is higher, which also applies to low resolution videos. It turns out that the peak to average ratio (PAR) of the frame sizes for videos with the same content but different resolutions are also similar. Thus, there is strong correlation between the PSNR and PAR.

It is interesting to note that for the low resolution video (i.e., 320×240), the Y-PSNR degrades smoothly as more P frames are employed. A perfect replica of the I frame can be obtained with a QP of 1. For high-resolution videos, the Y-PSNR degradation varies. This implies the quantization process may not be consistent for these videos, even though the VBR mode uses a fixed QP value for all encoded frames. A higher Y-PSNR variation tends to occur for the CBR videos. In the case, the QP value is changed depending on the content in order not to exceed a specific rate cap. The CBR video with a lower (or more aggressive) 1.5 Mbit/s rate cap shows the highest Y-PSNR variation, which is expected.

5.3.3   Limitations of PSNR

A video image that is displaced or rotated slightly (by say 1–2 samples or 1–2°) may lead to low PSNR but the subjective video quality may still be good. If there is a larger rotation of say 5–10°, the quality of the rotated video becomes unacceptable and the PSNR metric becomes reliable. Clearly, there is a quality threshold that must be determined. For example, is a 3° rotation acceptable compared to a 5° rotation? However, it must be pointed out that shifted or rotated video frames are rare with professional digital video capture and are not the result of codec or network transmission problems. Thus, if the decoded image is rotated, most likely, the source image is also rotated. This may happen when converting old film-based videos into digitized formats.

5.4   STRUCTURAL SIMILARITY INDEX

The structural similarity (SSIM) index [3] is a popular perceptual criterion for testing and optimizing video quality. SSIM accounts for higher-level structural information in the video content. Unlike error-based metrics such as PSNR, which can be inconsistent with human eye perception, SSIM takes into account the HVS. Some extensions include structural texture similarity index and color structural texture similarity index. SSIM is also commonly used as a method for testing the quality of lossy video encoders. For example, the popular open source H.264 encoder x264 is set by default to display an SSIM value at the end of an encoding operation. The SSIM index is a decimal value between 0 and 1. A value of 0 implies zero correlation with the original image whereas a 1 implies the exact same image as the original image. An SSIM value of 0.95 SSIM implies half as much variation from the original image as an SSIM value of 0.90.

There are three SSIM components. Luminance comparison in equation (5.6) is a function of the mean intensity. The contrast comparison in equation (5.7) is a function of variance. The structure comparison in equation (5.8) is function of covariance.

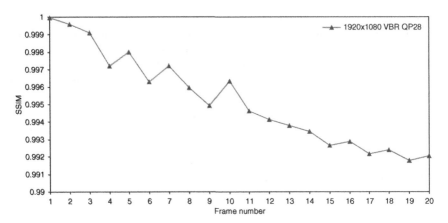

Figure 5.6   *Y*-SSIM.

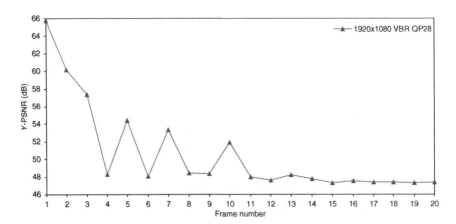

Figure 5.7   *Y*-PSNR.

Because these equations are fractions, the SSIM for the three components can be combined as a product in (5.9). This is in contrast to the weighted average for YUV-PSNR as defined in (5.5). As shown in Figures 5.6 and 5.7, there is a fairly strong correlation between *Y*-SSIM (a subjective metric) and *Y*-PSNR (an objective metric):

$$l(x, y) = \frac{2\mu_x\mu_y + C_1}{\mu_x^2 + \mu_y^2 + C_1} \tag{5.6}$$

$$c(x, y) = \frac{2\sigma_x\sigma_y + C_2}{\sigma_x^2 + \sigma_y^2 + C_2} \tag{5.7}$$

$$s(x, y) = \frac{\sigma_{xy} + C_3}{\sigma_x\sigma_y + C_3} \tag{5.8}$$

where μ_x and μ_y, and σ_x and σ_y, represent the mean and standard deviation of the decoded and original images, respectively:

$$\text{SSIM}(x, y) = l(x, y)^\alpha c(x, y)^\beta s(x, y)^\gamma \tag{5.9}$$

5.5 OBSERVABLE VERSUS PERCEPTUAL VISUAL ARTIFACTS

Observable or visible artifacts (e.g., choppy playback, frozen pictures, frame breakup) are more severe than perceptual visual artifacts (e.g., blurriness, blockiness) because they can generally be detected by viewers of all ages. Thus, unlike perceptual artifacts, the detection of observable artifacts is not prone to false alarms (unless the viewer has very poor eyesight). In Figure 5.8(b), the artifact is very obvious because it occurs on the subject and on the most important part of the subject – the eye. The background artifacts in Figure 5.8(c) are less visible or obvious. However, if temporal aspects

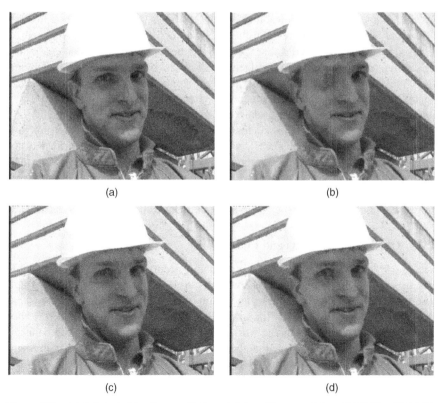

(a)   (b)

(c)   (d)

Figure 5.8 (a) Original video frame; (b) One visible artifact on subject, Y-PSNR = 35.03 dB, QP = 30; (c) Two visible artifacts on background, Y-PSNR = 35.10 dB, QP = 30; and (d) No visible artifact, Y-PSNR = 35.00 dB, QP = 32.

are accounted for by superposing the video frames in Figure 5.8(a) and (c), two artifacts become observable in the background. Note that since there is no movement
associated with the background, no change in the MVs is expected. In addition, the
Y-PSNR values of these frames are similar. Hence, evaluating the difference in the
Y-PSNR values of the two frames may not reveal the artifacts. Figure 5.8(d) shows
a uniform degradation in the overall quality of the video frame using a higher QP
value. In this case, there are no observable artifacts but the degradation may be perceptible depending on the visual response of the viewer. Since the Y-PSNR values
for Figure 5.8(b)–(d) are similar, clearly this objective video quality metric fails and
a subjective evaluation may be needed. However, if a higher Y-PSNR value (e.g.,
>40 dB) is employed, all observable artifacts can be detected.

5.5.1 Limited Information Provided by PSNR

Figure 5.8 illustrates a key limitation of objective video quality metrics such as PSNR.
Video images with identical content may give similar PSNR but the video quality
ranges from poor to good. PSNR does not indicate the number of observable artifacts
or the seriousness of these artifacts. This shows the inconsistency of the PSNR metric when assessing videos encoded with high QP values. Note that it is not possible
to recreate the same situation when the PSNR is high (e.g., >40 dB). This can be
achieved by using a lower QP value. Quantifying the degradation in video quality for
the different types of observable artifacts in Figure 5.8(b) and (c) is very challenging
(there may well be parts of the frame that are intentionally censored). Some video
streaming systems (e.g., adaptive streaming systems) adjust the QP value and resolution dynamically to prevent observable artifacts. For example, the video quality and
resolution is reduced as more packet losses are detected. However, the loss of picture
detail or fidelity may be visible for HD videos even though a lower resolution may not
result in an artifact. To summarize, the accuracy of practical measurement of video
quality may not be absolute or precise. It is also difficult to quantify such measurements. The overall picture quality can be poor even if there are no observable artifacts.
Conversely, the artifacts may be observable but the overall picture quality is good.

5.5.2 Observable Artifacts and Link Quality

Accurate video quality measurement or visual inspection of videos with low PSNR
may be unnecessary if these videos are known to be properly encoded. In general,
observable artifacts are closely associated with poor link quality. Thus, methods that
identify and fix other causes of the low PSNR assume a more important role. For
example, the link quality of the network can be assessed and measures can be taken
to improve the link quality. Note that these methods do not require the measurement
of video quality and can function independently.

5.5.3 Combined Spatial and Temporal Video Quality Assessment

Some video quality metrics attempt to assess the temporal and spatial aspects of a
video. For example, the ITU-T P.910 recommends the use of spatial and temporal

information based on the deviation of luminance samples of consecutive frames while other metrics evaluate the MVs in the encoded sequence.

5.6 ERROR CONCEALMENT

While detecting video degradation using various objective and perceptual metrics is useful, the ability to correct the visual artifacts to some degree assumes even greater importance. EC at the video decoder improves the overall video quality by correcting any observable or perceptible artifacts caused by errors associated with encoding, decoding, or transmission, which can be detected with the help of the syntax/semantics in each encoded video frame. Because EC can be applied to any decoded video bitstream, it is codec agnostic. Since EC targets the video bitstream rather than physical layer bitstream, its effectiveness is not limited to bit errors and random packet losses. In many cases, it can tolerate multiple (burst) packet losses that affect a single video frame. Thus, EC is more effective than forward error correction (FEC) at the physical layer. Like FEC, EC does not rely on channel feedback. However, unlike FEC and other bit-level error resilience methods, EC does not introduce any overhead bits although it may incur additional processing time. EC can be made adaptive to network conditions. For instance, EC is not required when sufficient bandwidth or a good channel is available. EC should be distinguished from postdecoding filtering such as deblocking filtering. Filters work on individual frames only and are designed to reduce blockiness in block-based coding rather than deal with transmission errors.

5.6.1 Error Resilience

Error resilience allows the blocks in a video frame to be reordered at the encoder. Figure 5.9 shows the impact of interlaced or dispersed slice groups. Just like bit

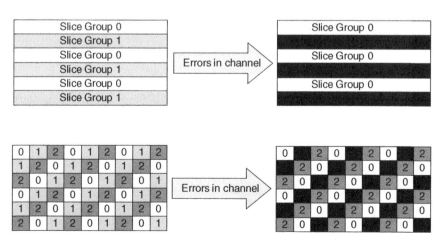

Figure 5.9 Spreading errors using multiple slice groups.

interleaving at the physical layer bitstream, error resilience scatters adjoining blocks throughout the frame, thereby improving error immunity and reducing the probability of a packet loss affecting a large region of the frame. In addition, video artifacts associated with packet losses can be concealed more effectively without relying on channel feedback or packet retransmission. Although the effectiveness of EC improves when combined with error resilience methods, error resilience is not mandatory for EC to function. In some cases, EC can be fairly effective when operated independently and therefore, additional overheads due to error resilience can be avoided. However, the use of error resilience methods mandates the use of EC, just as bit interleaving at the sender requires FEC to be applied at the receiver.

5.6.2 Impact on Visual Artifacts

EC changes the types of video artifacts. The frequency of occurrence of these artifacts becomes lower. Figure 5.10 illustrates the effectiveness of EC with and without error resilience. As can be seen, EC improves the video quality even without error resilience and is especially effective with minor errors affecting one or two

Figure 5.10 Impact of EC with and without error resilience.

blocks. However, EC may not prevent or reduce propagation errors. This is because frames corrected by EC may not serve as good references for future predicted frames.

5.6.3 Types of Error Concealment

EC schemes can be designed to handle partial or entire frame losses. Handling partial frame losses requires the detection of the corrupted or missing blocks after decoding in order to locate the affected areas of the image. One of the simplest methods is to replicate the damaged areas of a frame or a damaged frame from a previous frame. In the block copy method, each sample value of the concealed block(s) in a frame is copied from the corresponding sample of the previously decoded block(s). The concealed frame is displayed and stored in the reference frame buffer for decoding subsequent frames. In MV copy, the MVs and reference indices of the colocated blocks in the previously coded reference frames are copied and used to conceal the missing or damaged blocks. Thus, the corrupted frame is assumed to have the same motion as its reference frame. EC can also be based on spatial or temporal sample interpolation from the adjacent areas of the same frame or the reference frame.

5.6.4 Comparison of EC Methods

Copy methods are somewhat restrictive because they cannot protect the I frame or frames with scene changes. They may also be prone to error propagation and the performance depends on the scene complexity. Scenes with fast motion and rapid or frequent scene changes are difficult to conceal. Fortunately, other than high-action movie trailers and sports, most video content do not contain abrupt scene changes and fast motion occurring simultaneously.

Methods based on spatial interpolation do not suffer these disadvantages since the sample values of received or concealed neighboring blocks are derived from the same frame. Interpolation is employed when the frame to be concealed does not resemble the previous frame. However, the method may not always perform well. For I frames, a weighted average of the sample values of the surrounding blocks can be used. The process is applied to the luma and chroma components. In the case of P frames, motion compensated temporal interpolation can be performed to repair the damaged areas of the frame. Assuming that the motion is smooth and continuous in a frame, the MVs of the missing blocks are predicted from the surrounding decoded blocks.

5.6.5 Increasing Frame Rate Using EC

Many HDTVs with high video frame rates (e.g., 120, 240 Hz) employ temporal sample interpolation to increase the frame rate using videos with lower rates (e.g., 30 or 60 Hz). This is especially useful for high-action sports content. In doing so, the original video may reduce the storage space and transmission bandwidth requirements.

5.6.6 Actions Performed After EC

Actions performed after EC (e.g., deblocking filtering) may change the value of the samples of the block edges depending on the smoothness of the block edge and its neighbor. However, performing deblocking filtering with concealed blocks may potentially corrupt the samples of the correctly received block.

5.7 COLOR SCIENCE

Color science constitutes the psychological perception of color by the human eye and brain, the origin of color in materials, color theory in art, and the physics of electromagnetic radiation in the visible range. It is also known as chromatics, chromatography, or colorimetry. Color science is key to improving the video quality of original videos. Professional videos are crisp and clear because they captured with the best equipment that takes into account the distribution of light power versus color wavelength and the interaction with spectral sensitivities of the eye's receptors. Color science is also key to producing professional animated movies.

5.7.1 Color Reception

The ability of the human eye to distinguish colors is based on the sensitivity of different cells in the retina to different wavelengths. Because perception of color stems from the varying spectral sensitivity of different types of cone cells in the retina to different parts of the spectrum, colors may be defined and quantified by the degree to which they stimulate these color receptor cells or cones. The human eye is trichromatic. Hence, the retina contains three types of cones. Short wavelength cones are most responsive to blue. Middle-wavelength and long-wavelength cones are sensitive to green and red respectively. Light is reduced to these three color components by the eye. The three types of cones give three signals based on the degree of stimulation. It has been estimated that the human eye can distinguish roughly 10 million different colors. Another light-sensitive cell in the eye is called the rod. When light is bright enough to stimulate the cones, rods play virtually no role in vision at all. On the other hand, in dim light, the cones are understimulated, leaving only the signal from the rods, which lead to a colorless response. If some color-sensing cones are missing or less responsive to incoming light, that person can distinguish fewer colors and is said to be color deficient.

5.7.2 Color Reproduction

Colors may vary in several different ways, including hue, saturation, brightness, and gloss. The dominant wavelength is largely responsible for hue and refers to a single wavelength of light that produces a response most similar to the light source. Two different light spectra (e.g., white light from fluorescent light and sunlight) that have the same effect on the three cones in the human eye will be perceived as the same color.

In general, a mixture of colors (e.g., red and green) cannot produce a response that is identical to a spectral (or fully saturated) color such as yellow. Fortunately, natural scenes rarely contain spectral colors. Thus, such scenes can usually be approximated well by color-mixture systems. Another problem with color reproduction systems is associated with the capture devices. The color sensors in these devices are very different from the receptors in the human eye. Color acquisition by these devices can be poor if there is unusual lighting in the scene. For color information captured in digital form, color management techniques can help to avoid distortions. For example, such techniques may map the input colors into a color palette that can be reproduced with minimum distortion.

REFERENCES

1. Human visual system, http://webvision.med.utah.edu.
2. ITU's tutorial on objective perceptual assessment of video quality, http://www.itu.int/ITU-T/studygroups/com09/docs/tutorial_opavc.pdf.
3. Z. Wang, A. Bovik, H. Sheikh, and E. Simoncelli, "Image Quality Assessment: From Error Visibility to Structural Similarity," IEEE Transactions on Image Processing, Vol. 13, No. 4, April 2004, pp. 600–612.

HOMEWORK PROBLEMS

5.1. If separate rate-distortion curves for the luma and chroma components are employed, this will lead to three different average bit rates, one for each component. Explain whether this method of separating the measurements is better than combining the luma and chroma components (as in YUV-PSNR) or evaluating only the luma component (as in Y-PSNR).

5.2. How can the video quality of 3D videos be assessed effectively? Is SSIM adequate in addressing MV corruption, including disparity vector corruption?

5.3. Justify the following statement: a high PSNR provides a reliable measure of good video quality whereas a low PSNR may not always result in poor video quality. Can PSNR detect motion artifacts in successive frames? Given the limitations of PSNR, should it be replaced by "better" video quality metrics?

5.4. In perceptual coding, quantization noise is shaped so that it is masked by the signal energy. The technique has been applied successfully in voice/audio compression. Explain whether a similar technique can be used to improve video quality.

5.5. Which of the following videos (with the same content) will produce a higher PSNR? (a) a coded 720p video or (b) a 1080p video coded to a bit rate that is equivalent to the coded 720p video.

5.6. How effective is a video quality meter that simply flags an artifact whenever a damaged or missing MB is detected compared to a meter that tries to identify the type of artifact (e.g., blockiness, missing MV).

5.7. Identical video content coded with the same QP value but with different resolutions may lead to a similar PSNR and PAR of the frame sizes. Will a larger frame size that causes a higher PAR also leads to a higher PSNR? Take into account the fact that both PSNR and PAR may not always be consistent.

5.8. In general, it is more difficult for a lossy encoder to code a video with fine details, including videos captured at high resolutions. This may in turn lead to a lower PSNR because the decoded video may not be a good reproduction of the original. Refine this observation in the context of Figure 5.5. Which factor has a more dominant influence on PSNR: QP or video resolution?

5.9. Is constant PSNR achievable for a video coded with a fixed QP value?

5.10. Watch a sports video such as a tennis match and compare the video quality of 60 and 240 Hz HDTVs when viewed at 2 and 15 ft. What can you conclude about increasing the frame rate and viewing distance with respect to video quality? Visual artifacts tend to occur in high-action sports videos but they may not be perceptible. Would a slow-motion playback reveal these artifacts more succinctly?

5.11. Although the HVS works with a continuous flow of photons, it is useful to analyze the maximum video frame rate of the HVS. Suppose a tennis player serves at 140 mph on a 1080p video. Compute the minimum frame rate in order to avoid any visible artifact. Note that when a badminton shuttlecock is served at 200 mph, the player can still retrieve it.

5.12. Justify the following statement: "Great video quality with HD/UHD may not be as important as good choice in video content."

6

CODING PERFORMANCE OF H.262, H.264, AND H.265

This chapter provides experimental results on the coding performance of three generations of video coding standards that were produced by the MPEG/VCEG working groups. By reviewing the file-based and frame-based coding performance, we provide an in-depth comparison of the H.265/HEVC, H.264/AVC, and H.262/MPEG-2 standards. We believe that this is the first time the frame-based video coding performance has been reported. This offers useful insights into the coding efficiencies of the individual frames, which are either intracoded or intercoded, and how they respond to different types of video content. For example, the use of unidirectionally predicted frames contributes significantly to the high H.265 coding efficiency for complex video content whereas bidirectionally predicted frames are far less effective for such content. On the other hand, P frames may be coded more efficiently than B frames by H.262 for certain types of high-resolution content. The coding efficiency of the individual frame types allows the frequency of the different frame types to be optimized in a group of pictures (GOP). We also provide extensive results detailing the impact of new H.265 features on video coding efficiency. Specifically, we studied the impact of changing the size of the coding and transform units on the coding efficiency and encoding time. The factors affecting the encoder complexity and resilience are also investigated. Overall, the optimization of the video coding efficiency requires a painstaking effort in reviewing the coding performance for a wide variety of video content under different coding parameter settings and video quality.

Next-Generation Video Coding and Streaming, First Edition. Benny Bing.
© 2015 John Wiley & Sons, Inc. Published 2015 by John Wiley & Sons, Inc.

Table 6.1   H.265 Video Coding Configurations.

Name	Configuration
main	Uses InternalBitDepth of 8
he10	Uses InternalBitDepth of 10
intra_main, intra_he10	All frames are I frames
lowdelay_P_main, lowdelay_P_he10	Uses an I frame followed by P frames GOP size is 4
lowdelay_main, lowdelay_he10	Uses an I frame followed by B frames. GOP size is 4
randomaccess_main, randomaccess_he10	An I frame is inserted every 32 frames. All other frames are B frames. GOP size is 8

6.1   CODING PARAMETERS

The HEVC Test Model (HM) reference software [1] provides a test platform for normative features of H.265. Like the H.264 JM software [2], the HM software is not optimized for commercial use due to the long encoding and decoding delays. There are eight standard video coding configurations that are employed by the HM software. They are listed in Table 6.1. The key parameters are shown in Table 6.2. The HM software supports frame rates of up to 300 Hz. When the encoder operates on a 10-bit video in the 8-bit mode, prior to encoding, each 10-bit sample (x) is converted to an 8-bit value $(x + 2)/4$ clipped to the [0, 255] range. Similarly, when the encoder operates on an 8-bit video in the 10-bit mode, each 8-bit sample (y) is converted to a 10-bit value $4y$. This conversion facility is built into the reference HM encoder.

Figure 6.1 shows the coding efficiencies of the eight HM configurations. As expected, the intracoded configurations are the least efficient. The random access configurations produce the smallest compressed file size but the difference with the low delay configurations is marginal. Figure 6.2 shows the coding efficiencies for the high-efficiency configurations that employ GOP sizes of 4 or 8. Again, the random access configuration is the most efficient due to the larger GOP size. The difference in the coded file size for the first two configurations (i.e., low delay_P_he10 and lowdelay_he10) is marginal for *Park Scene*. As will be shown later, the coding efficiency of P and B frames are similar due to the low content complexity of this video, thus leading to comparable coded frame sizes.

6.1.1   Coding Block Size

Figure 6.3 shows the impact of reducing the luma coding block size from the maximum allowed size of 64×64 to 32×32 or 16×16 samples using the H.265 main profile. In general, a smaller block size degrades the H.265 compression efficiency. The degradation is more significant for 16×16 compared to 32×32 blocks as well as for higher-resolution videos. The higher degradation for 16×16 blocks not only

Table 6.2   Key Parameter Settings for H.265 Configurations.

H.265 Parameter	intra_main, intra_he10	lowdelay_P_main, lowdelay_P_he10, lowdelay_main, lowdelay_he10	randomaccess_main, randomaccess_he10
MaxCUWidth	64	64	64
MaxCUHeight	64	64	64
MaxPartitionDepth	4	4	4
QuadtreeTULog2MaxSize	5	5	5
QuadtreeTULog2MinSize	2	2	2
QuadtreeTUMaxDepthInter	3	3	3
QuadtreeTUMaxDepthIntra	3	3	3
GOP Size	1	4	8
Intra Period	1	−1*	32
QP	32	32	32
InternalBitDepth	8 (main), 10 (he10)	8 (main), 10 (he10)	8 (main), 10 (he10)
SAO	1	1	1
ALF, DB	1	1	1
AMP	1	1	1
TS	1	1	1

*Only the first frame in the GOP is an I frame.

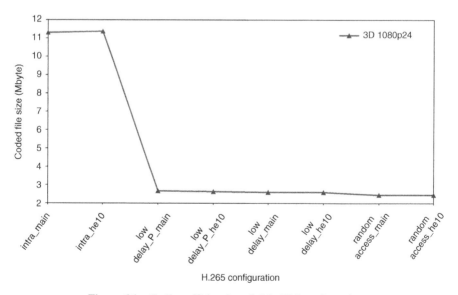

Figure 6.1   Coding efficiencies of eight HM configurations.

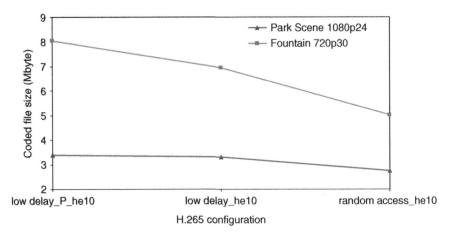

Figure 6.2   Coding efficiencies of high-fficiency (HE) configurations.

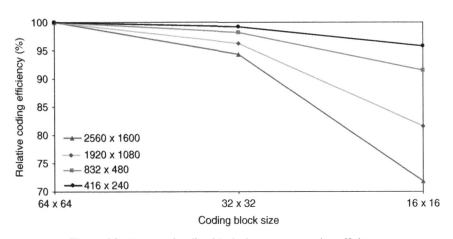

Figure 6.3   Impact of coding block size on compression efficiency.

demonstrates the lower efficiency of using a smaller block but also the limited flexibility when the block partition depth is reduced from 4 (for 64×64) to 2 (for 16×16). Clearly, a larger block size is more desirable for encoding high-resolution videos, including high definition (HD) and ultra-high definition (UHD) videos.

As shown in Figure 6.4, a benefit of using smaller block sizes is the faster encoding. Although there are more blocks to process, smaller blocks allow simpler and faster intraprediction, interprediction, and quadtree splitting, thereby shortening the encoding time. A reverse trend occurs for the decoding time. In this case, a smaller block size of 16×16 increases the decoding time quite substantially due to the need to buffer larger compressed videos, a direct result of the lower encoding efficiency. Since user devices such as smartphones and tablets are limited in processing power,

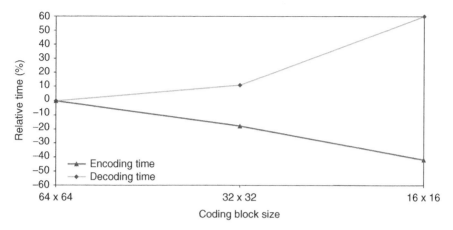

Figure 6.4 Impact of coding block size on encoding and decoding time.

using a larger block size may be more desirable in terms of a smaller compressed video and shorter decoding time. More cache memory may be required for intermediate storage but the overall memory requirement for storing and decoding each frame is lower due to the superior coding efficiency.

6.1.2 Transform Block Size

Larger coding block structures impact MV prediction, in particular, uniform motion, and should be complemented by larger transform blocks. Figure 6.5 shows the impact of reducing the transform block size from the maximum allowed size of 32×32 to 16×16 or 8×8 samples using the H.265 main profile. Although the trend is similar

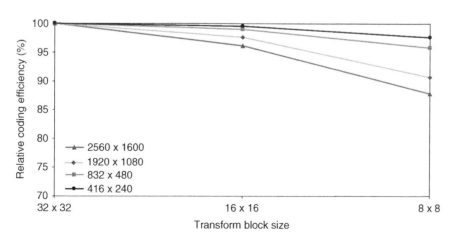

Figure 6.5 Impact of transform block size on compression efficiency.

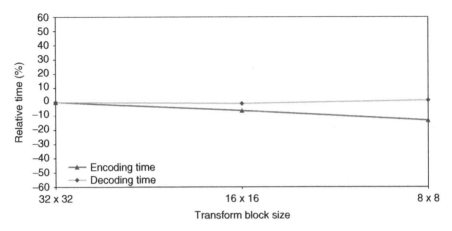

Figure 6.6   Impact of transform block size on encoding and decoding time.

to the coding blocks, the reduction in coding efficiency is smaller because constraining the maximum coding block size indirectly constrains the maximum transform block size but the converse is not true. The smaller reduction in the coding efficiency shows that the benefits of using larger coding blocks are not simply due to the larger transforms. The chroma components tend to be more sensitive to changes in the transform block size than the luma component. H.265 allows the transform block size in a coding block to be selected independently of the prediction block size (with few exceptions). This is controlled via the residual quadtree (RQT), which has a selectable depth. As evident in Figure 6.6, there is hardly any impact on the decoding time when different transform block sizes are used. However, there are some gains (about 10% maximum) when a smaller transform block size is used for encoding.

6.1.3   TMVP, SAO, AMP

In H.265, several syntax elements allow various tools to be configured or enabled. Among these are parameters that specify the minimum and maximum coding and transform block size, and transform hierarchy depth. If a fixed coding block size and QP are employed, a marginal bit rate improvement of 1–3% can be achieved when disabling temporal motion vector prediction (TMVP) and sample adaptive offset (SAO). TMVP predicts the MV of a current PU using a colocated PU in the reference frame. In practice, a large block size or higher QP may be used to offset the TMVP and SAO overheads. SAO has a larger impact on the subjective quality than on the Y-PSNR. These tools have a marginal impact on encoding or decoding time. When the asymmetric motion partitioning (AMP) tool is disabled, a bit rate improvement of about 1% can be achieved. The significant increase in encoding time can be attributed to the additional motion search and decision that is needed for AMP. Disabling the TS tool does not change the coding efficiency for most video content. The TS tool is most effective for content such as computer graphics with camera captured content and animated movies. For such content, disabling of the TS tool increases the bit rate savings by 6–7%.

6.2   COMPARISON OF H.265 AND H.264

We evaluate the coding performance of H.265 and H.264 and assess the strengths, weaknesses, and relative gains. It will be shown that motion level (e.g., moderate vs low motion), content type (e.g., high vs low complexity), and video resolution (e.g., 1080p vs 720p) may impact the performance of the codecs. We employ two HD videos for most of the tests. *Fountain* is a professionally captured 720p (30 Hz) video that contains highly complex content with low motion. *Park Scene* is a 1080p (24 Hz) video with low complexity and moderate motion that has been used as a test sequence by MPEG/VCEG. There are 240 frames in each of these videos. Although these videos are not representative of popular commercial videos (e.g., full-motion sports, high-action movies), the tests do offer useful insights and may be extended to a broad variety of content types as well as 4K UHD videos.

The videos are encoded in the VBR format where rate control is disabled and a fixed QP value is used to code every video frame. The use of a lower QP value will lead to higher quality coded videos whereas the converse is true for higher QP values. The P frames are configured as unipredicted referenced frames. The B frames are configured as unreferenced frames and require bidirectional prediction. For a consistent comparison, we employ the JM and HM reference software for coding and decoding. Parallelism (available in H.265) has not been enabled as this may impact the coding efficiency and encoding time.

6.2.1   Absolute Coding Efficiency

Some interesting trends can be made by analyzing the absolute and relative coding gains of H.264 and H.265. The absolute coding gain relates to the actual compressed H.264 or H.265 video. A larger compressed video corresponds to a lower coding gain. The relative coding gain refers to the difference in the size of a compressed H.264 video and a compressed H.265 video (i.e., the bit rate savings of H.265 over H.264). For each data point, the average coded file size using three QP values is obtained. Figures 6.7 and 6.8 illustrate the absolute coding gains of H.264 and H.265. Although the resolution of the *Fountain* video (720p) is lower than the *Park Scene* video (1080p), the absolute H.264 and H.265 coding efficiencies for *Fountain* are lower than *Park Scene* for a broad range of QP values (i.e., 16–34). In other words, the compressed videos for *Fountain* are larger than *Park Scene*. The higher complexity of the *Fountain* video is primarily responsible for the larger compressed videos. The converse is true for very low QP values (i.e., 8–10). The higher resolution of *Park Scene* reduces the absolute coding efficiency of H.264 and H.265 for very low QP values (i.e., 8–10) when compared to *Fountain*. This suggests the H.264 and H.265 encoders become less efficient when coding high-resolution videos at very high quality (i.e., at very low QP values) and this is independent of content complexity.

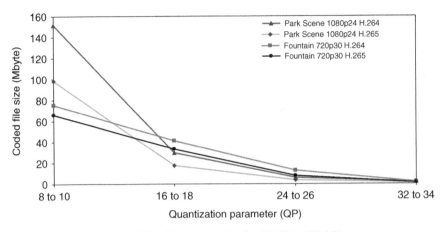

Figure 6.7   Compressed size for H.265 and H.264.

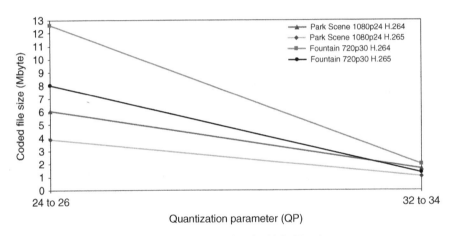

Figure 6.8   Expanded view for high QP values.

6.2.2   Relative Coding Gain

Figure 6.9 shows the relative coding gains for the two videos. The relative coding gain grows when smaller QP values are chosen. The relative gains are quite consistent for the *Park Scene* video (about 36% on the average), even though the absolute coding gain is lower for low QP values. This can be explained as follows. For low QP values, the larger relative coding gain for this video is offset by a larger compressed video, hence the ratio of the relative coding gain over the compressed file size becomes similar to ratios derived for higher QP values. However, for the *Fountain* video, the relative coding gain increases at a much slower rate as the QP value decreases. In this case, the smaller relative coding gain compared to the larger compressed file size causes

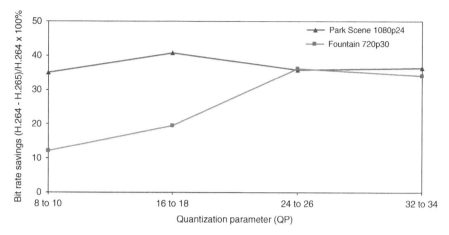

Figure 6.9   Bit rate savings for H.265 over H.264.

the ratio to decrease as the compressed video becomes larger when low QP values are employed. Overall, the results indicate that for low quality video compression (i.e., using high QP values), the relative coding gains of H.265 over H.264 are quite stable and somewhat independent of the content type, resolution, and frame rate. On the other hand, for high-quality compression (i.e., using low QP values), high content complexity and resolution may degrade the coding performance of both H.264 and H.265.

6.2.3   Videos with Different Levels of Motion

Figure 6.10 shows relative coding gains (in terms of relative bit rate savings) of using the H.265 main profile over the H.264 high profile for two 10-s 1080p videos (24 Hz) with different levels of motion. Because the YUV-PSNR metric is reliable and consistent only for high video quality (i.e., YUV-PSNR ≥ 40 dB), the video quality for lower YUV-PSNR values may vary. There are several interesting trends. First, the efficiency gap between H.265 and H.264 narrows for higher motion content (in spite of H.265 employing a much greater number of directions for intraprediction and more partitions for interprediction). Second, the actual improvement varies with the video quality. For low video quality (e.g., YUV-PSNR < 35 dB), higher gains can be achieved. For high video quality, the gains tend to level off. Thus, higher gains may require a sacrifice in quality. Third, the gains are more consistent for YUV-PSNR values greater than 40 dB. Thus, the 40 dB threshold should be used to evaluate the coding efficiencies of high-quality video content. The higher YUV-PSNR also implies a lower QP is used, which reduces the relative coding gains due to the larger compressed videos. In this case, lower information loss is caused by the video encoding process.

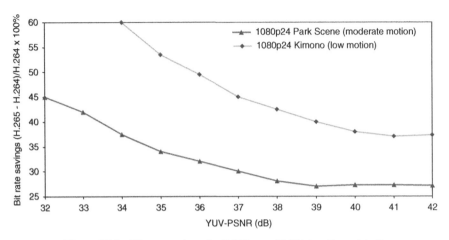

Figure 6.10   Bit rate savings for H.265 over H.264 (motion-related).

6.3   FRAME CODING COMPARISON

The overall compressed file size is a useful indicator of the coding efficiency of the video coding standard. To fully understand the efficiency of the coding mechanisms, an evaluation of the coding characteristics of the individual frame (i.e., I, P, and B frames) is required. To this end, six different H.265 configurations to evaluate the coding performance of each frame type are employed. These configurations are listed in Table 6.3. The H.265 main profile and the H.264 high profile are used for encoding. By maintaining the same QP value of 28, a consistent comparison of the relative coding gain between the two standards can be achieved.

Configurations 1 and 2 test the impact of using two different bit depths (i.e., 8 bit and 10 bit). Configurations 2–4 are used to compare the impact of decreasing the coding block size from 64×64 to 32×32 to 16×16. Configurations 4 and 5 are used to evaluate the impact of reducing the transform block size from 32×32 to 8×8 whereas the coding block size is fixed at 16×16. Finally, configurations 5 and 6 are used to evaluate the additional bit overheads for activating the new advanced features of H.265 such as SAO, AMP, and TMVP. There are flags to turn these tools on or off. By setting these parameters, the contribution of these tools to the coding improvements of H.265 can be assessed. Adaptive loop filter (ALF) is a third in-loop filtering step after SAO filtering and deblocking filtering (DBF). ALF improves video quality with diamond-shape 2D filtering. The filter coefficients are signaled in the bitstream and can therefore be designed based on image content and distortion of the reconstructed frame. The filter restores the reconstructed frame by minimizing the MSE between the source and reconstructed frames. With the full set of loop filtering stages (i.e., DBF, SAO, ALF), the decoding complexity may become comparable to motion compensation. Thus, although ALF was considered during the development of HEVC, it was not included in final standard. LMChroma is an intracoding method that uses the linear model (LM) to predict the chroma signal based on the

Table 6.3   Parameter Settings for H.265 Configurations 1–6.

H.265 Parameter	1	2	3	4	5	6
MaxCUWidth	64	64	32	16	16	16
MaxCUHeight	64	64	32	16	16	16
MaxPartitionDepth	4	4	3	2	2	2
QuadtreeTULog2MaxSize	5	5	5	5	3	3
QuadtreeTULog2MinSize	2	2	2	2	2	2
QuadtreeTUMaxDepthInter	3	3	3	3	1	1
QuadtreeTUMaxDepthIntra	3	3	3	3	1	1
Intra period	32	32	32	32	32	32
GOP size	8	8	8	8	8	8
QP	28	28	28	28	28	28
InternalBitDepth	10	8	8	8	8	8
SAO	1	1	1	1	1	0
ALF, DB	1	1	1	1	1	0
LMChroma	1	1	1	1	1	0
NSQT	1	1	1	1	1	0
AMP	1	1	1	1	1	0
TS	1	1	1	1	1	0
TMVPMode	1	1	1	1	1	0

reconstructed luma signal. nonsquare quad tree (NSQT) is an extension of RQT that enables rectangular TUs. By using nonsquare transforms, asymmetric motion compensated predictor partitioning can be supported. It is important to emphasize that the results and findings in the later sections require these tools to be turned on.

6.3.1   I Frame Coding Efficiency, Quality, and Time

Figure 6.11 shows the relative I frame coding gains of H.265 over H.264. The relative gains for all configurations are comparatively lower than the compressed videos (9–13% vs 36% for *Park Scene*). This can be attributed to the lower efficiency in intracoding and the larger size of the compressed I frame. The YUV-PSNR of the coded I frames for H.265 and H.264 are closely matched in Figure 6.12. This provides a basis for comparing the bit rate savings between the H.265 and H.264 I frames. The I frame encoding times for H.265 are similar or faster than H.264 (Figure 6.13). Clearly, the use of larger coding blocks (e.g., 64×64) and the corresponding larger partition depth (e.g., 4), larger transform blocks (e.g., 32×32), together with the advanced H.265 features (e.g., SAO, AMP, TMVP) all help to improve the coding efficiency and encoding time of H.265 over H.264.

The magnitude of the relative coding gain for H.265 is higher for *Park Scene* (14% for configuration 2) than *Fountain* (8% for configuration 2). However, the relative gain for each video when activating different H.265 features is not significant (<5% for *Park Scene*, <1% for *Fountain*). For example, by comparing the gains for *Park Scene* when configurations 2 and 6 were applied, the difference is about 4%, which

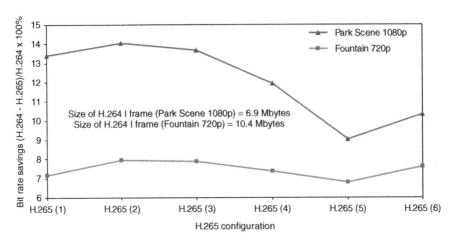

Figure 6.11   I frame coding efficiency gain for H.265 over H.264.

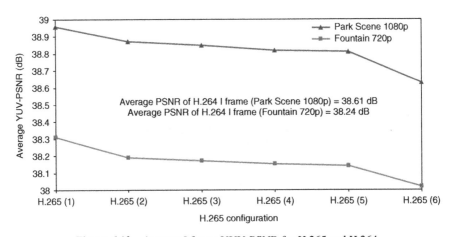

Figure 6.12   Average I frame YUV-PSNR for H.265 and H.264.

is small given the vast differences in the parameter settings (e.g., 16×16 vs 64×64 coding blocks, use of smaller transform blocks, and disabling of all advanced features for configuration 6). This difference becomes almost 0% for the *Fountain* video. This suggests that different combinations of the H.265 features may not improve intracoding efficiency significantly, especially for complex video content.

There is marginal or no difference in the relative coding gain, YUV-PSNR, and encoding time for 8 and 10 bit depths. A 10 bit depth incurs a slightly higher amount of bit overheads than an 8 bit depth. As a result, a slight degradation in the relative coding gain for 10 bit depth is observed. Disabling the advanced H.265 features gives the fastest encoding but the poorest YUV-PSNR. Because bit overheads are reduced when the advanced features are deactivated, the relative coding gain becomes slightly

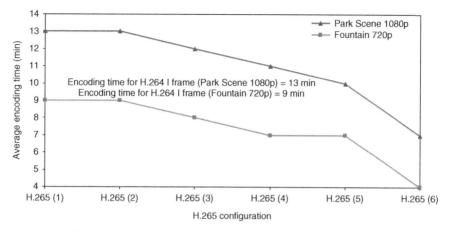

Figure 6.13   Average I frame encoding time for H.265 and H.264.

better than the case when these features were activated (configuration 5). Turning off the advanced features reduces the overheads by about 1–2% for I frames.

6.3.2   P Frame Coding Efficiency, Quality, and Time

Figure 6.14 shows the relative P frame coding gains for six different H.265 config-urations. The relative gains are comparable to the compressed videos. However, the absolute gains are lower than I frames. This can be attributed to the higher efficiency in unidirectional interprediction (uniprediction), leading to a smaller size for the coded P frames. The YUV-PSNR values of the coded P frames for H.265 and H.264 are

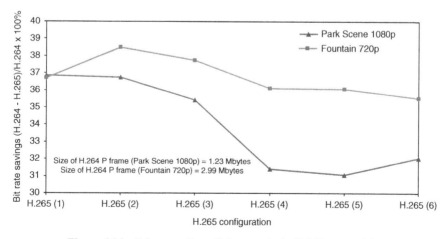

Figure 6.14   P frame coding efficiency gain for H.265 over H.264.

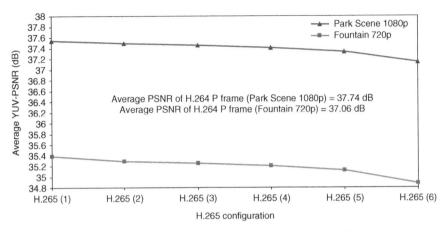

Figure 6.15   Average P frame YUV-PSNR for H.265 and H.264.

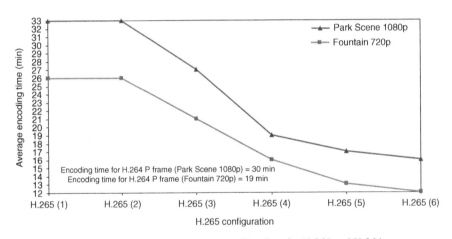

Figure 6.16   Average P frame encoding time for H.265 and H.264.

closely matched in Figure 6.15. This provides a basis for comparing the bit rate savings between the H.265 and H.264P frames. The P frame encoding times for H.265
can be slower or faster than H.264 (Figure 6.16).

Unlike the case for I frames, the magnitude of the relative coding gain for H.265
is lower for *Park Scene* (36.8% for configuration 2) than *Fountain* (38.5% for configuration 2). In addition, the *Park Scene* video shows a clear dip in the coding gain
whereas there is no dip for the *Fountain* video when the advanced features are deactivated. This suggests that the advanced features improve the uniprediction of videos
with complex content. Complex videos such as *Fountain* also benefit from smaller
coding blocks. Thus, the variation in the coding gain is small when the size of the
coding blocks is modified. Like the I frames, the relative improvement for each video

when activating different H.265 features is not significant (<6% for *Park Scene*, <2% for *Fountain*). Again, this shows that the coding efficiency for the P frames is less sensitive to changes in the H.265 parameter settings for complex video content.

There is marginal or no difference in the relative coding gain, YUV-PSNR, and encoding time for 8 and 10 bit depths. Consistent with the I frames, a slight degradation in the relative coding gain for 10 bit depth is observed. Disabling the advanced H.265 features leads to the fastest encoding but the poorest YUV-PSNR. Again, the reduced bit overheads slightly improve the relative coding gain over the case when these features were activated (configuration 5). Turning off the advanced features reduces the overheads by about 1% for the P frames of *Park Scene*. There are no additional overheads for the *Fountain* video.

6.3.3   B Frame Coding Efficiency, Quality, and Time

Figure 6.17 shows the relative B frame coding gains for six different H.265 configurations. The relative gains are lower than the P frames. For the *Fountain* video containing complex content, the relative gains can turn negative. In this case, H.264 coding of B frames becomes more efficient than H.265 although the difference is not significant (<3%). This suggests that the new H.265 features are ineffective for bidirectional interprediction of videos with complex content. The YUV-PSNR values of the coded B frames for H.265 and H.264 are closely matched in Figure 6.18. This provides a basis for comparing the bit rate savings between H.265 and H.264 B frames. The encoding times for H.265 can be slower or faster than H.264 (Figure 6.19).

Like the case for I frames, the magnitude of the relative coding gain for H.265 is much higher for *Park Scene* (17.4% for configuration 2) than *Fountain* (0.7% for configuration 2). Both videos show a dip in the coding gain when the advanced features are deactivated. Like the I frames, the relative improvement for each video when activating different H.265 features is not significant (<7% for *Park Scene*, <4% for

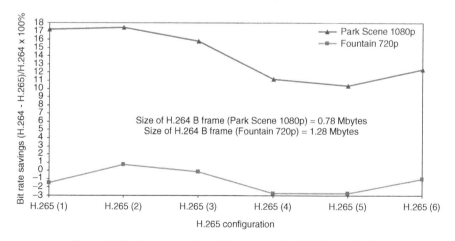

Figure 6.17   B frame coding efficiency gain for H.265 over H.264.

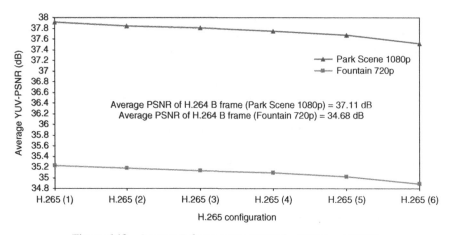

Figure 6.18   Average B frame YUV-PSNR for H.265 and H.264.

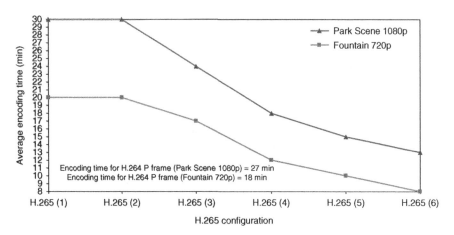

Figure 6.19   Average B frame encoding time for H.265 and H.264.

Fountain). This shows that the coding gain for the B frames is less sensitive to changes in the H.265 parameter settings for complex video content such as the *Fountain* video.

There is marginal or no difference in the relative coding gain, YUV-PSNR, and encoding time for 8 and 10 bit depths. A slight degradation in the relative coding gain for 10 bit depth is observed. Disabling the advanced H.265 features leads to the fastest encoding and a slight improvement in the relative coding gain over the case when these features were activated (configuration 5). However, this is done at the expense of the poorest YUV-PSNR. Turning off the advanced features reduces the overheads by about 2–3% for the B frames.

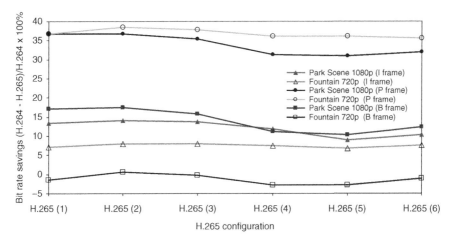

Figure 6.20 Frame coding efficiency gain for H.265 over H.264.

6.3.4 Overall Frame Coding Efficiency, Quality, and Time

Additional insights can be obtained by combining the overall relative coding gains for each frame type. As depicted in Figure 6.20, the absolute gains are highest for the I frames whereas the relative gains are highest for the P frames. High complexity content benefit most from unipredicted P frames, achieving the best relative coding gain among all frame types and content types. In addition, the gains are quite consistent even as the H.265 parameter settings are varied and are not dependent on the use of advanced H.265 features. Like the I frames, the relative coding gains for B frames tend to be substantially lower than P frames. For complex video content, the coding efficiency of the H.265 B frames may become marginally lower than H.264. The coded frame sizes are shown in Figure 6.21. The variation in the coded sizes for all frame types is marginal, which suggests the coded frame size is somewhat independent of the H.265 configuration. The size of the coded B frames is smaller than the P frames for both videos. The difference is marginal for the *Park Scene* video but higher for the *Fountain* video due to the higher complexity. The size of the I frame is on the order of 5 or 6 times larger than the P and B frames.

As expected and shown in Figure 6.22, the YUV-PSNR values of the intracoded I frames are the best compared to the other frame types. This is independent of the content type and resolution. If all other parameter settings are the same, an 8 bit depth gives a slightly higher relative coding gain than a 10 bit depth (compare configurations 1 and 2). The amount of bit overhead is dependent on the content type. A more complex video increases the bit overheads (0.1–0.2% for *Park Scene* vs ~2% for *Fountain*). A 10 bit depth gives the best YUV-PSNR but achieves a very marginal improvement (<0.1%) than an 8 bit depth.

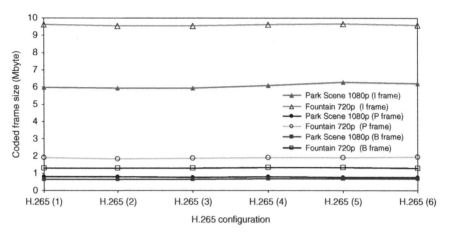

Figure 6.21   Coded H.265 frame sizes.

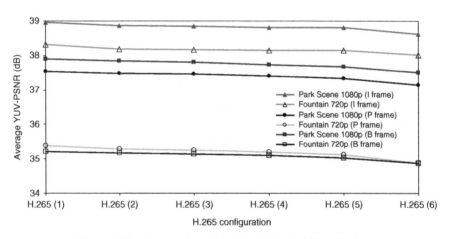

Figure 6.22   Average frame YUV-PSNR for H.265 and H.264.

For complex content (such as the *Fountain* video), the intercoded B and P frames exhibit the lowest quality (i.e., lowest YUV-PSNR values). In this case, the YUV-PSNR values for the P and B frames are almost identical, which implies that the video quality remains unaffected by the combination of P and B frames that are employed in a GOP. The video resolution plays a key role in the encoding time (Figure 6.23). A higher video resolution leads to longer intercoding delay regardless of content type. P frames require longer encoding times than B frames. On the other hand, intracoding achieves the lowest encoding delay. In addition, a lower video resolution leads to a lower intracoding delay and this is independent of the content type. Thus, the I frames for the 720p *Fountain* video require the least time to encode among all frame types. Disabling the advanced features give the fastest

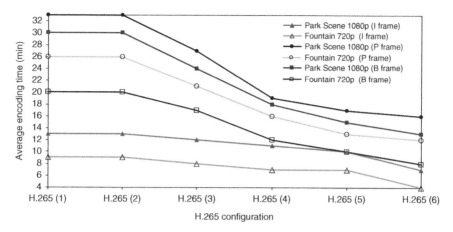

Figure 6.23   Average frame encoding time for H.265 and H.264.

frame encoding time (~50% reduction compared to the case when the features were activated). There is no change in the encoding time for 8 and 10 bit depths, as seen in configurations 1 and 2. A reduction in the coding block size and partition depth clearly reduces the encoding time. A more thorough analysis on the impact of the coding and transform block size is provided in the following sections.

6.4 IMPACT OF CODING BLOCK SIZE ON FRAME CODING EFFICIENCY

Figure 6.24 shows how the size of the coding block impacts the relative coding efficiency for different frame types when compared to a coding block size of 64×64. Clearly, a smaller coding block size reduces the frame coding efficiency. The I frames are least affected by the choice of the coding block size and achieves better coding efficiency than the P and B frames. The maximum degradation in the intracoding efficiency for 16×16 blocks when compared to 64×64 blocks ranges from 1% to 2%. A higher video resolution (i.e., *Park Scene*) leads to a lower intracoding efficiency. Similarly, the video resolution impacts the intercoding efficiency. In this case, the maximum degradation in the intercoding efficiency for 16×16 blocks when compared to 64×64 blocks ranges from 3% to 9%. The coding efficiency for P frames is slightly worse than for B frames. Additional work is reported in [4, 5].

6.4.1 Impact of Transform Block Size on Frame Coding Efficiency

Figure 6.25 shows how the size of the transform block size impacts the relative coding efficiency for different frame types when compared to a transform block size of 32×32. A smaller size causes zero or marginal reduction in the frame coding efficiency. The maximum degradation in the coding efficiency for 8×8 transform

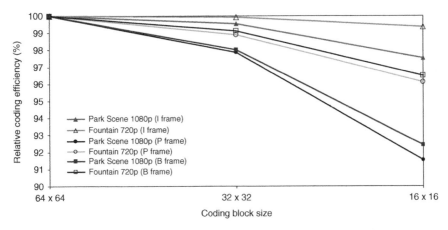

Figure 6.24   H.265 coding efficiency for different coding block size.

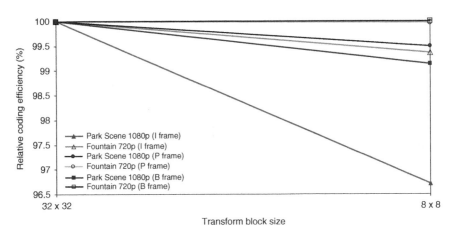

Figure 6.25   H.265 coding efficiency for different transform block size.

blocks when compared to 32×32 transform block ranges from 0% to just over 3%. The I frames tend to be more sensitive to changes in the transform block size, with higher resolution I frames suffering more degradation. The intercoded P and B frames of *Fountain* maintain the coding efficiency even as the size of the transform block decreases. The results show that constraining the maximum coding block size indirectly constrains the maximum transform block size but the converse is not true. The same conclusion can also be deduced from Figure 6.20, which shows very marginal variation in the coding efficiency for virtually all frame types using configurations 4 and 5 (with the exception of the intracoded *Park Scene* I frame). This is in contrast to the other configurations (i.e., 1, 2, 3, 6). Recall that configuration 4 employs a coding block size of 16×16 and a bigger transform block size of 32×32 (with a transform

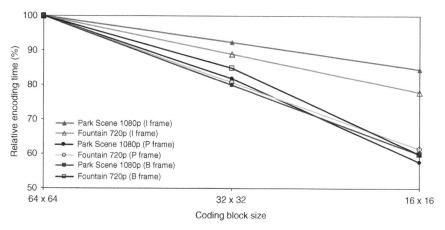

Figure 6.26 H.265 encoding time for different coding block size.

partition depth of 3) whereas configuration 5 uses a coding block size of 16×16 and a smaller transform block size of 8×8 (with a transform partition depth of 1).

6.4.2 Impact of Coding Block Size on Frame Encoding Time

Figure 6.26 shows how the size of the coding block impacts the relative frame encoding time when compared to a coding block size of 64×64. Clearly, a smaller coding block size reduces the frame encoding time. Intracoding achieves the smallest reduction in encoding time when smaller blocks are chosen. The decrease in the encoding time for the higher resolution *Park Scene* I frame is less than that for *Fountain* I frame when a smaller block size is chosen. This suggests that video resolution rather than content complexity impact the intracoding time for smaller blocks. The reduction in the encoding time is substantial (roughly 40%) for intercoded frames (i.e., B and P frames) when smaller blocks are selected. This reduction is roughly similar for the intercoded frames.

6.4.3 Impact of Transform Block Size on Frame Encoding Time

Figure 6.27 shows how the size of the transform block impacts the relative frame encoding time when compared to a transform block size of 32×32. A smaller block size causes 0% or up to 18% reduction in the frame encoding time. The complexity of the video plays a key role, with no changes in the encoding time for the I frames of *Fountain*. However, reduction of about 16–18% in the encoding time is observed for the P and B frames of *Fountain*.

6.4.4 Impact of CU Size on Encoding Time

Figure 6.28 shows the impact of a fixed CU size on the total computational time during encoding. The computational time relates to the complexity in the encoding

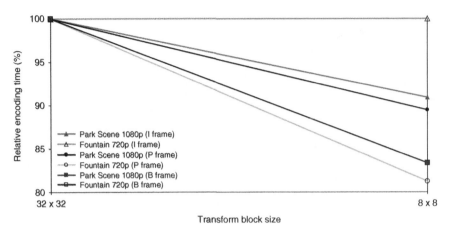

Figure 6.27   H.265 encoding time for different transform block size.

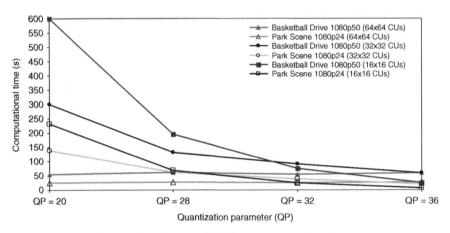

Figure 6.28   Impact of CU size on computational time.

because it arises from determining whether a CU should be split to meet a given rate distortion requirement. For example, a big CU size may generate high distortion for an image region with fine-grained features. CU sizes of 64×64, 32×32, and 16×16 were tested. For low QP values, 16×16 CUs lead to longer computational time than 64×64 CUs. In addition, high-motion content requires higher computational time, regardless of the QP value or CU size. Hence, a low QP value coupled with high-motion content may increase the computational time for smaller CUs. This is evident for the *Basketball Drive* video that is encoded with a QP value of 20. In this case, the computational time increases 12 times when the CU size is reduced from 64×64 to 16×16. For high QP values, 16×16 CUs may incur shorter computational time than 32×32 or 64×64 CUs. This applies to both *Basketball Drive* and

Park Scene videos and can be attributed to the low quality videos. For 64×64 CUs, the computational time is fairly consistent over a broad range of QP values. Thus, the computational time for big CUs is somewhat insensitive to changes in the QP value. Overall, the results show that if deeper quadtree spitting is employed, which may lead to smaller CU sizes, the computational time for high-quality encoding may become significant, especially with high-motion content. In other words, larger CUs should be employed for high-quality encoding whenever possible. Conversely, for low quality encoding, CU partitioning may be unnecessary and the smallest CU size should be chosen for all content types to minimize computational time.

6.4.5   Decoding Time

We wish to assess the computational time for various decoding components of H.265. The evaluation provides insights to the optimization of the decoder functions for multiscreen display platforms such as PCs, laptops, tablets, and smartphones. Several 1080p videos were decoded using the intra_main and randomaccess_main configurations without activating parallelism. The results were averaged over these videos. For intracoded frames, since the majority of the coded bits represent transform coefficients, the parsing of these coefficients represents a key decoding bottleneck. The next most demanding task is intraframe prediction due to the high number of prediction directions. This decoding time is significantly higher than randomaccess_main because there are more intracoded frames for the intra_main configuration. For random access, motion estimation takes up a significant portion of the decoding time (over 40%). Specifically, the interpolation filtering process in the motion compensation is time consuming. The combined filtering of DBF and SAO accounts for about 20% of the decoding time, which is comparable to intra_main. CABAC decoding takes up 25%, about 10% higher than intra_main. Although H.265 employs large transforms and SAO, they do not appear to significantly affect the decoding time (Figure 6.29).

6.5   SUMMARY OF CODING PERFORMANCE

We summarize the key observations of our evaluations in Tables 6.4–6.11. Using bidirectionally predicted B frames for coding complex content may lead to weaker coding gains than H.264 and the poorest video quality among all frame types. This may be attributed to the improved H.265 motion compensated predictions that lead to highly uncorrelated prediction residuals. This in turn degrades the performance of DCT, which may not be able to supply a compact representation of the residuals with few significant coefficients.

6.6   ERROR RESILIENCY COMPARISON OF H.264 AND H.265

It is common to use the packet loss metric to quantify the error resiliency of a packetized network. For example, if 1 packet is lost for every 1000 packets transmitted,

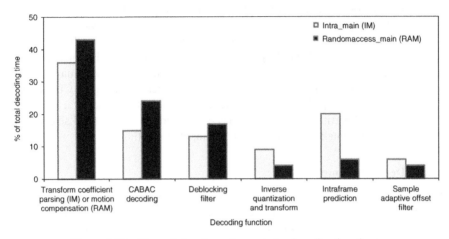

Figure 6.29   Computational requirements for decoding functions.

Table 6.4   Impact of QP.

Absolute coding efficiency	Reduces for content with high complexity for a broad range of QP values (e.g., 16–34). Reduces for content with high resolution for very low QP values (e.g., 8–10)
Relative coding gain (compared to H.264)	Reduces for content with high complexity and low QP values. Fairly stable for high QP values, independent of content type, resolution, and frame rate

Table 6.5   H.265 Frame Coding Characteristics.

Frame Type	Relative Coding Gain	YUV-PSNR	Encoding Time
I	Substantially lower than P frames (>20%)	Better than P and B frames (1–3 dB), independent of content type and resolution	Shorter than P and B frames, similar or faster than H.264
P	Better than I and B frames, independent of content type and resolution		Longer than I and B frames, slower or faster than H.264
B	Substantially lower than P frames (>20%)		Slower or faster than H.264

Table 6.6   Impact of Higher Content Complexity.

Frame Type	Relative Coding Gain	YUV-PSNR
I	Moderate decrease (<7%). Lower variation	Slight decrease (<0.8 dB)
P	Increases (<5%). Resilient to changes in parameters. Lower variation	Decreases (<3 dB), almost identical to B frames
B	Substantial decrease (<17%). Slightly worse than H.264. Lower variation	Decreases (<3 dB), almost identical to P frames

Table 6.7   Impact of Higher Video Resolution.

Frame Type	Encoding Time
I	Increases, independent of content type
P	Increases, independent of content type
B	Increases, independent of content type

Table 6.8   Impact of Increased Bit Depth (i.e., 10 bit).

Frame Type	Relative Coding Gain	YUV-PSNR	Encoding Time
I	Slight decrease (<0.7%)	Very marginal increase (<0.1%)	Unaffected
P	Slight decrease (<2%)*	Very marginal increase (<0.1%)	Unaffected
B	Slight decrease (<2%)*	Very marginal increase (<0.1%)	Unaffected

*Higher decrease of ~2% for more complex content.

Table 6.9   Impact of Activating H.265 Advanced Features.

Frame Type	Relative Coding Gain	YUV-PSNR	Encoding Time
I	Slight decrease (<2%)	Very marginal increase (<0.2%)	Higher increase than P and B frames
P	Resilient: unchanged for complex content, slight decrease (~ 1%) for less complex content	Very marginal increase (<0.2%)	Lower increase than I and B frames
B	Slight decrease (2–3%)	Very marginal increase (<0.2%)	

Table 6.10   Impact of Using Smaller Coding Blocks (e.g., 64×64 vs 16×16).

Absolute coding efficiency	Decreases for intracoded and intercoded frames
	Slight decrease for intracoded frames ($<2.2\%$)
	Higher decrease for intercoded frames ($<8.5\%$)
	Degradation for P frames slightly worse than B frames
Video resolution	Higher resolution degrades relative intracoding and intercoding gain
Encoding time	Substantial decrease for intercoded frames ($\sim40\%$)
	Smaller decrease for intracoded frames (12–20%)
	Affected by video resolution rather than content complexity
	Increases computational time

Table 6.11   Impact of Using Smaller Transform Blocks (e.g., 32×32 vs 8×8).

Absolute coding efficiency	Zero or marginal decrease ($\sim3.3\%$)
	Intracoded frames more sensitive to changes in transform block size
Video resolution	Higher resolution degrades relative intracoding gain
Content complexity	Slight decrease in relative coding gain for intracoded frames ($<0.6\%$). Intercoded frames are unaffected
Encoding time	0–18% decrease
	Intracoded frames of complex content are unaffected

then the packet loss rate of the network is 1/1000 or 0.001. In order to maintain an acceptable level of quality of service, keeping a low packet loss rate is a major goal of many networks. If such a network is used to carry digitized video, the packet loss metric will ultimately have to be traced back to the frame type in order to quantify the error resiliency of the video coding standard. This is because a single video frame is typically transported by several packets. Hence, depending on the location of the error, a single video frame may corrupted by one or more packets. The error resiliency of H.264 and H.265 can be quantified in terms of the frame type (i.e., I, P, B frame types). Entire frames or portions of the frames can be removed from a video sequence and the sequence is decoded. By examining the JM/HM decoding messages, the error resiliency of H.264 and H.265 can be deduced and quantified.

6.6.1   H.264 Error Resiliency

The following JM decoding message shows the impact of removing an IDR frame. The decoding is aborted due to the missing I frame.

```
------------------------------------------------------------
Input H.264 bitstream : fountain.264
Output decoded YUV    : test.yuv
Input reference file  :
------------------------------------------------------------
no: 0, nal_unit_type: 7, len:80
no: 1, nal_unit_type: 8, len:32
no: 2, nal_unit_type: 8, len:32
no: 3, nal_unit_type: 8, len:32
no: 4, nal_unit_type: 5, len:1491048
Image Format : 1280x720 (1280x720)
Color Format : 4:2:0 (8:8:8)
------------------------------------------------------------
POC must = frame# or field# for SNRs to be correct
------------------------------------------------------------
Frame  POC Pic# QP SnrY SnrU SnrV Y:U:V Time(ms)
------------------------------------------------------------
no: 5, nal_unit_type: 1, len:1487216
00000(IDR) 0 0 28 4:2:0 50
no: 6, nal_unit_type: 1, len:1505416
no: 7, nal_unit_type: 1, len:1486528
00001( I ) 2 1 28 4:2:0 51
An unintentional loss of pictures occurs! Exit
```

The following JM decoding message shows the impact of removing a P frame. The decoding is aborted due to the missing P frame.

```
------------------------------------------------------------
Input H.264 bitstream : fountain.264
Output decoded YUV    : test.yuv
Input reference file  :
------------------------------------------------------------
no: 0, nal_unit_type: 7, len:80
no: 1, nal_unit_type: 8, len:32
no: 2, nal_unit_type: 8, len:32
no: 3, nal_unit_type: 8, len:32
no: 4, nal_unit_type: 5, len:1520544
Image Format : 1280x720 (1280x720)
Color Format : 4:2:0 (8:8:8)
------------------------------------------------------------
POC must = frame# or field# for SNRs to be correct
------------------------------------------------------------
Frame POC Pic# QP SnrY SnrU SnrV Y:U:V Time(ms)
------------------------------------------------------------
```

```
no: 5, nal_unit_type: 1, len:332112
00000(IDR)  0   0 28 4:2:0 51
no: 6, nal_unit_type: 1, len:430640
00001( P ) 2 1 31 4:2:0 26
no: 7, nal_unit_type: 1, len:312416
00002( P ) 4 2 30 4:2:0 31
no: 8, nal_unit_type: 1, len:527080
no: 9, nal_unit_type: 1, len:350800
00003( P ) 6  3 31 4:2:0 27
An unintentional loss of pictures occurs! Exit
```

The following JM decoding message shows the impact of removing an unreferenced B frame. The decoding continues in spite of the missing B frame. Thus, H.264 is insensitive to losses of unreferenced B frames.

```
------------------------------------------------------------
Input H.264 bitstream : fountain.264
Output decoded YUV    : test.yuv
Input reference file  :
------------------------------------------------------------
no: 1, nal_unit_type: 7, len:80
no: 2, nal_unit_type: 8, len:32
no: 3, nal_unit_type: 8, len:32
no: 4, nal_unit_type: 8, len:32
no: 5, nal_unit_type: 5, len:1570920
Image Format : 1280x720 (1280x720)
Color Format : 4:2:0 (8:8:8)
------------------------------------------------------------
POC must = frame# or field# for SNRs to be correct
------------------------------------------------------------
Frame POC Pic# QP SnrY SnrU SnrV Y:U:V Time(ms)
------------------------------------------------------------
no: 6, nal_unit_type: 9, len:8
no: 7, nal_unit_type: 1, len:731016
00000(IDR) 0 0 28 4:2:0 52
no: 8, nal_unit_type: 9, len:8
no: 9, nal_unit_type: 1, len:208592
00008( B ) 16 1 29 4:2:0 48
no:10, nal_unit_type: 9, len:8
no:11, nal_unit_type: 1, len:92608
00004( B ) 8 2 30 4:2:0 29
no:12, nal_unit_type: 9, len:8
no:13, nal_unit_type: 1, len:33048
no:14, nal_unit_type: 9, len:8
no:15, nal_unit_type: 1, len:33344
```

```
00002( B ) 4 3 31 4:2:0 24
no:16, nal_unit_type: 9, len:8
no:17, nal_unit_type: 1, len:112568
00003( b ) 6 4 32 4:2:0 19
no:18, nal_unit_type: 9, len:8
no:19, nal_unit_type: 1, len:31376
00006( B ) 12  4 31 4:2:0  24
no:20, nal_unit_type: 9, len:8
no:21, nal_unit_type: 1, len:35944
00005( b ) 10 5 32 4:2:0 17
no:22, nal_unit_type: 9, len:8
no:23, nal_unit_type: 1, len:678320
00007( b ) 14 5 32 4:2:0 18
no:24, nal_unit_type: 9, len:8
no:25, nal_unit_type: 1, len:213672
00016( B ) 32 5 29 4:2:0 44
no:26, nal_unit_type: 9, len:8
no:27, nal_unit_type: 1, len:106696
00012( B ) 24 6 30 4:2:0 29
no:28, nal_unit_type: 9, len:8
no:29, nal_unit_type: 1, len:38312
00010( B ) 20 7 31 4:2:0 24
no:30, nal_unit_type: 9, len:8
no:31, nal_unit_type: 1, len:33048
00009( b ) 18 8 32 4:2:0 20
no:32, nal_unit_type: 9, len:8
no:33, nal_unit_type: 1, len:119576
00011( b ) 22 8 32 4:2:0 20
no:34, nal_unit_type: 9, len:8
no:35, nal_unit_type: 1, len:37376
00014( B ) 28 8 31 4:2:0 25
no:36, nal_unit_type: 9, len:8
no:37, nal_unit_type: 1, len:36832
00013( b ) 26 9 32 4:2:0 20
```

The following JM decoding message shows the impact of removing a referenced B frame. The decoding is aborted due to the missing B frame.

```
------------------------------------------------------------
Input H.264 bitstream : fountain.264
Output decoded YUV    : test.yuv
Input reference file  :
------------------------------------------------------------
no: 0, nal_unit_type: 9, len:8
no: 1, nal_unit_type: 7, len:80
```

```
no: 2, nal_unit_type: 8, len:32
no: 3, nal_unit_type: 8, len:32
no: 4, nal_unit_type: 8, len:32
no: 5, nal_unit_type: 5, len:1570920
Image Format : 1280x720 (1280x720)
Color Format : 4:2:0 (8:8:8)
------------------------------------------------------
POC must = frame# or field# for SNRs to be correct
------------------------------------------------------
Frame POC Pic#  QP SnrY SnrU SnrV Y:U:V Time(ms)
------------------------------------------------------
no: 6, nal_unit_type: 9, len:8
no: 7, nal_unit_type: 1, len:731016
no: 8, nal_unit_type: 9, len:8
no: 9, nal_unit_type: 1, len:208592
00000(IDR) 0 0 28 4:2:0 53
An unintentional loss of pictures occurs! Exit
```

6.6.2 H.265 Error Resiliency

The following HM decoding message shows the impact of removing an I frame. The decoding is aborted due to the missing I frame.

```
HM software: Decoder Version [7.1rc1][Linux][GCC 4.6.3]
[64 bit]
no: 0, TYPE:25, bits:32
no: 1, TYPE:26, bits:160
no: 2, TYPE:27, bits:48
TDecCavlc::parsePPS(): m_bUseWeightPred=0 m_
uiBiPredIdc=0
no: 3, TYPE:28, bits:680
no: 4, TYPE: 8, bits:1387712
no: 5, TYPE:28, bits:832
no: 6, TYPE: 1, bits:1380160
no: 7, TYPE: 1, bits:1380160
TComInputBitstream::readByte(UInt&): Assertion `m_fifo_
idx < m_fifo->size()' failed.
POC 0 TId: 0 ( I-SLICE, QP 28 ) [DT 0.100] [L0 ] [L1 ]
Aborted (core dumped)
```

The following HM decoding message shows the impact of removing a P frame. The decoding is aborted due to the missing P frame.

```
HM software: Decoder Version [7.1rc1][Linux][GCC 4.6.3]
[64 bit]
```

```
no: 0, TYPE:25, bits:40
no: 1, TYPE:26, bits:296
no: 2, TYPE:27, bits:48
TDecCavlc::parsePPS(): m_bUseWeightPred=0 m_
uiBiPredIdc=0
no: 3, TYPE:28, bits:632
no: 4, TYPE: 8, bits:1421416
no: 5, TYPE:28, bits:576
no: 6, TYPE: 1, bits:111984
no: 7, TYPE: 1, bits:111984
no: 8, TYPE:28, bits:472
no: 9, TYPE: 1, bits:206416
POC  0 TId: 0 ( I-SLICE, QP 28 ) [DT 0.100] [L0 ] [L1 ]
no:10, TYPE: 1, bits:206416
TComInputBitstream::readByte(UInt&): Assertion 'm_fifo_
idx < m_fifo->size()' failed.
POC 1 TId: 0 ( P-SLICE, QP 31 ) [DT 0.040] [L0 0 ] [L1 ]
Aborted (core dumped)
```

The following HM decoding message shows the impact of removing an unreferenced B frame. The decoding is aborted due to the missing B frame.

```
HM software: Decoder Version [7.1rc1][Linux][GCC 4.6.3]
[64 bit]
no: 0, TYPE:25, bits:40
no: 1, TYPE:26, bits:312
no: 2, TYPE:27, bits:48
TDecCavlc::parsePPS(): m_bUseWeightPred=0
m_uiBiPredIdc=0
no: 3, TYPE:28, bits:864
no: 4, TYPE: 8, bits:1481728
no: 5, TYPE:28, bits:584
no: 6, TYPE: 1, bits:790144
no: 7, TYPE: 1, bits:790144
no: 8, TYPE:28, bits:440
no: 9, TYPE: 1, bits:278536
POC 0 TId: 0 ( I-SLICE, QP 28 ) [DT 0.120] [L0 ] [L1 ]
no:10, TYPE: 1, bits:278536
no:11, TYPE:28, bits:296
no:12, TYPE: 1, bits:138256
POC 8 TId: 0 ( B-SLICE, QP 29 ) [DT 0.090] [L0 0 ]
[L1 0 ]
no:13, TYPE: 1, bits:138256
TComInputBitstream::readByte(UInt&): Assertion 'm_fifo_
idx < m_fifo->size()' failed.
```

```
POC  4 TId: 0 ( B-SLICE, QP 30 ) [DT 0.040] [L0 0 8 ]
[L1 8 0 ] Aborted (core dumped)
```

The following HM decoding message shows the impact of removing a referenced B frame. The decoding is aborted due to the missing B frame.

```
HM software: Decoder Version [7.1rc1][Linux][GCC 4.6.3]
[64 bit]
no: 0, TYPE:25, bits:40
no: 1, TYPE:26, bits:312
no: 2, TYPE:27, bits:48
TDecCavlc::parsePPS(): m_bUseWeightPred=0
m_uiBiPredIdc=0
no: 3, TYPE:28, bits:864
no: 4, TYPE: 8, bits:1481728
no: 5, TYPE:28, bits:584
no: 6, TYPE: 1, bits:790144
no: 7, TYPE: 1, bits:790144
no: 8, TYPE:28, bits:440
no: 9, TYPE: 1, bits:278592
POC  0 TId: 0 ( I-SLICE, QP 28 ) [DT 0.120] [L0 ] [L1 ]
no:10, TYPE: 1, bits:278592
TComInputBitstream::readByte(UInt&): Assertion 'm_fifo_
idx < m_fifo->size()' failed.
POC  8 TId: 0 ( B-SLICE, QP 29 ) [DT 0.090] [L0 0 ]
[L1 0 ] Aborted (core dumped)
```

Thus, H.265 is sensitive to losses of unreferenced and referenced frames.

6.7 H.264/H.265 VERSUS H.262

Unlike H.262, there are higher computational demands in H.264/H.265 decoding due to the smaller block size and in-loop DBF (Figure 6.30). In general, H.264 decoder is about 2 times more complex than the MPEG-4 Advanced Simple Profile (ASP) decoder (used in DivX and Xvid). The H.264 encoder is about 10 times as complex as the MPEG-4 ASP coder. In H.262, an 8×8 DCT is used to transform the predicted error blocks from the spatial domain to the frequency domain. A comparison of the profile, application, decoder complexity, and efficiency of H.264 and H.262 is shown in Table 6.12.

6.7.1 Performance Comparison

We employ 10 HD videos, including four 3D videos to perform the evaluation. The H.265 main profile (configuration 2), H.264 high profile, and H.262 encoder from [3]

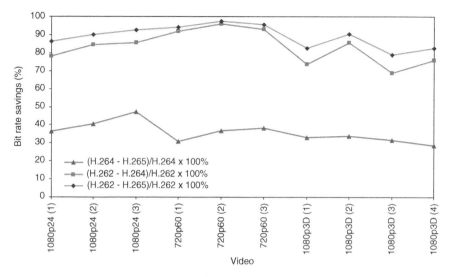

Figure 6.30   Bit rate savings for H.262, H.264, and H.265.

Table 6.12   H.264 Versus H.262.

Profile	Typical Application	Additional Decoder Complexity over H.262	Typical Efficiency over H.262
Baseline	Low delay applications	2.5 times	1.5 times
Extended	Mobile streaming	3.5 times	1.75 times
Main	Broadcast video	4 times	2 times

are employed. As depicted in Figure 6.30, the relative coding gains are roughly 40% for H.265 over H.264 and 85% for H.264 over H.262. It is challenging to streamline the parameters for all three video coding standards. For example, both H.264 and H.265 allow 52 quantization levels whereas only 32 levels are allowed in H.262. To maintain a consistent YUV-PSNR (Figure 6.31), we set $QP = 32$ for H.264 and $QP = 34$ for H.265. Figure 6.32 shows that the H.262 encoding time is far shorter than H.264 and H.265. The H.264 encoding time is in turn, shorter than H.265 in most cases.

6.7.2   H.262 Frame Coding Efficiency

We now examine the frame coding efficiency of H.262. As shown in Figure 6.33, the H.262 I frame coding efficiency for *Fountain* is poorer than *Park Scene*, even though *Fountain* is a lower resolution video. The lower efficiency is due to the higher content complexity in *Fountain*. Figure 6.34 shows a similar trend for the P frames. In this

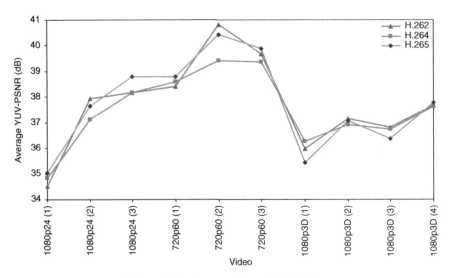

Figure 6.31   Average frame YUV-PSNR.

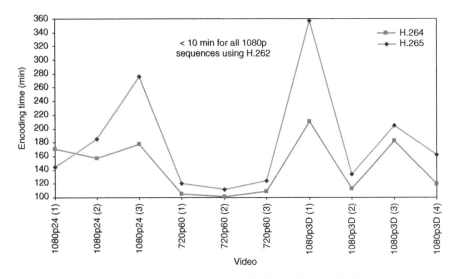

Figure 6.32   Encoding time for H.262, H.264, and H.265.

case, the coding efficiency for *Park Scene* over *Fountain* is more than doubled for a broad range of QP values. The trend is repeated for the B frames (Figure 6.35). In this case, we evaluate the coding performance when 2B frames and 4B frames are employed for every P frame. It turns out that there is very marginal difference

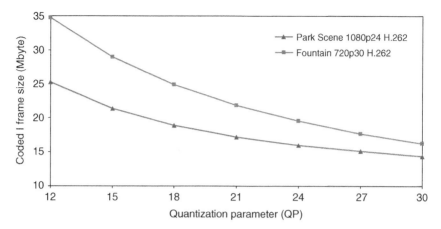

Figure 6.33   I frame coding efficiency for H.262.

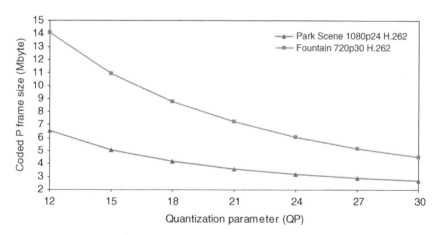

Figure 6.34   P frame coding efficiency for H.262.

in the coding efficiency when the number of B frames is increased from 2 to 4. By combining Figures 6.34 and 6.35, we obtain Figure 6.36. For *Park Scene*, the best coding efficiency is obtained for P frames. The P frame coding efficiency can be better than the B frame coding efficiency. Note that in Figure 6.21, the P frame coding efficiency for H.265 is poorer than the B frame coding efficiency for the same video (i.e., *Park Scene*), which is more typical. For *Fountain*, the best coding efficiency is obtained for B frames, which is consistent with the trend for H.265 in Figure 6.21. This suggests that bidirectionally predicted B frames are useful for improving the H.262 coding efficiency of complex content. However, the difference in the H.262

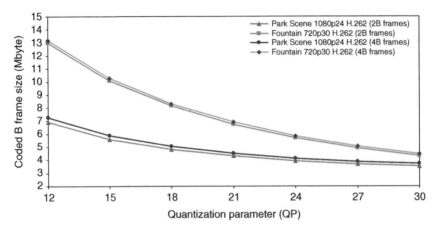

Figure 6.35 B frame coding efficiency for H.262.

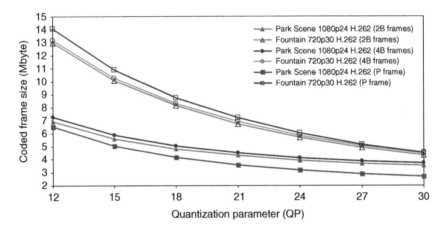

Figure 6.36 P frame and B frame coding efficiency for H.262.

coding efficiency for P and B frames (i.e., intercoded frames) is insignificant when the encoded video contains complex content.

6.7.3 Impact of GOP Size

Figure 6.37 shows the P frame coding efficiency when different GOP sizes are chosen. The QP value is fixed at 21. Clearly, a bigger GOP size more efficient but the gain levels off as bigger GOPs are chosen. GOP sizes of 10 or 12 are recommended. The use of bigger GOPs requires more buffering and increases the encoding and decoding delay. To summarize, the H.262 coding efficiency is dependent on the use of P and B frames, GOP size, and content.

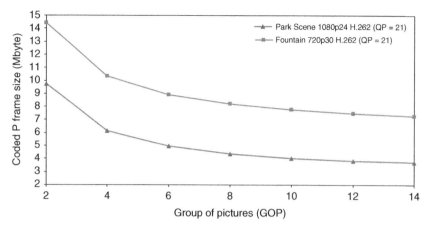

Figure 6.37 P frame coding efficiency for H.262.

REFERENCES

1. H.265/HEVC Test Model (HM) Reference Software, http://hevc.hhi.fraunhofer.de/HM-doc.
2. H.264/AVC Joint Model (JM) Reference Software, http://iphome.hhi.de/suehring/tml.
3. H.262/MPEG-2 encoder, http://mpeg.org.
4. J. Ohm, G. Sullivan, H. Schwarz, T. K. Tan, and T. Wiegand, "Comparison of the Coding Efficiency of Video Coding Standards—Including High Efficiency Video Coding (HEVC) ," IEEE Transactions on Circuits and Systems for Video Technology, Vol. 22, No. 12, December 2012, pp. 1669–1684.
5. F. Bossen, B. Bross, K. Suhring, and D. Flynn, "HEVC Complexity and Implementation Analysis," IEEE Transactions on Circuits and Systems for Video Technology, Vol. 22, No. 12, December 2012, pp. 1685–1696.

HOMEWORK PROBLEMS

6.1. Intracoded frames typically provide the best quality after they are decoded, they also suffer from a lower coding efficiency than intercoded frames, often by a few orders of magnitude. The use of intracoded frames should therefore be minimized. Unfortunately, intracoded frames are normally inserted at the start of a GOP, even when the video content does not require it. Devise a method to insert intracoded frames in a sequence of video frames efficiently. Evaluate the drawbacks for using the new method.

6.2. Intercoded frames require references, which can be unidirectionally predicted (e.g., P frames) using one past reference frame or bidirectionally predicted (e.g., B frames) using two reference frames (one past frame and one future frame). The bipredicted frames are normally assumed to offer the best coding efficiency. Hence, many video encoders tend to use more bipredicted frames than unipredicted frames in a GOP to improve the coding efficiency. However, this assumption is not always valid depending on the content and the video coding standard. In addition, the number of unipredicted and bipredicted frames that make up a fixed GOP pattern is often preset in advance and cannot be changed to match scene variations in the video content, thus compromising the overall coding efficiency. In some cases, intercoded frames are used as a substitute for scenes that should be intracoded in order to conform to the GOP structure. Consequently, the intercoded frames become very large, further degrading the coding efficiency. Devise a method to insert intercoded frames in a GOP efficiently. Evaluate the advantages and drawbacks of the new method.

6.3. Provide a list of advantages for using bigger CUs.

6.4. Design an algorithm to adapt the number of P and B frames in a GOP according to the video coding standard to obtain the best overall coding efficiency.

6.5. Using the results from Section 6.3, estimate the maximum and minimum buffer size to store the coded *Park Scene* and *Fountain* video frames for a two-second GOP.

7

3D VIDEO CODING

3D movies, especially animated movies, have enjoyed some measure of success. Autostereoscopic (glass-free) displays may be ready to move 3D entertainment viewing up a new level. Multiview 3D video bitstreams contain much redundancy yet they are sometimes coded separately (e.g., left-eye and right-eye views) to improve parallelism (i.e., simultaneous coding of multiple views with no degradation in spatial or temporal resolution). This increases the transmission bandwidth and storage requirements, especially for supporting wide-angle coverage involving many simultaneous views. This chapter focuses on the coding of multiple 3D views for efficient storage and network transport.

7.1 INTRODUCTION

The simplest form of 3D TV is a stereoscopic 3D (S3D) TV, where the display is capable of rendering only two views, one for each eye, so that the scene is perceived as 3D when wearing appropriate eye glasses. Thus, such displays project stereoscopic image pairs to the viewer. The source video captures stereo pairs in a two-view setup, with cameras mounted side by side and separated by the same distance as between a person's eyes. The active eye glasses turn dark over one eye for about 8 ms while showing the other eye an image and then vice versa for the complementary image. The S3D displays normally employ infrared signaling to activate and synchronize

Next-Generation Video Coding and Streaming, First Edition. Benny Bing.
© 2015 John Wiley & Sons, Inc. Published 2015 by John Wiley & Sons, Inc.

to the glasses. More sophisticated autostereoscopic displays are able to render and display more views in such a way that the perceived 3D video depends on the user's location with respect to the TV. For example, the user can move the head to see what is behind a certain object in a scene. This feature brings a new form of interactivity and an immersive feeling to TV viewers that have never been experienced before. In addition, no eye glasses are required. Since multiple views of the video sequences are captured by different cameras in different positions and angles, they are just different representations of the same scene. Thus, redundancy in the view direction exists for 3D video although there is dependency between the views.

7.1.1 3D Video Transmission and Coding

3D videos require two or more views to present stereoscopic depth. Active-shutter TVs and Blu-ray players employ frame-sequential or simulcast (one channel per eye) for local 3D playback. This requires the video decoder to operate twice as fast to support two 3D views or images. In addition, using this method for 3D transmission will require at least double the bandwidth for standard 2D transmission.

7.1.2 View Multiplexing

Temporal or spatial multiplexing of the views can be used to deliver 3D services without increasing bandwidth, as shown in Figures 7.1 and 7.2. In temporal multiplexing, the full spatial resolution of each view is preserved at the expense of reduced temporal resolution. Upconversion to the original video frame rate may be required at the receiver. Because of the reduced frame rate, full-resolution temporal multiplexing tends to require a lower bandwidth than spatial multiplexing. In addition, the original video frame is retained in temporal multiplexing and this may allow more efficient coding than spatial multiplexing, which requires the video frame to be subsampled. Other important advantages of temporal multiplexing include minimum postprocessing (e.g., no view upconversion) for playback and simplified preprocessing (e.g., no view packing or downconversion). Many 3D TVs and HDMI interfaces support temporal multiplexing with two full-resolution views occupying each frame.

Spatial or frame-compatible multiplexing preserves the temporal resolution but reduces the spatial resolution. View packing within a video frame is a necessary

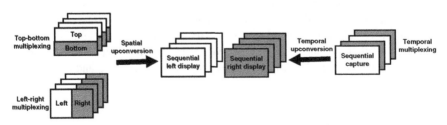

Figure 7.1 Spatial and temporal view multiplexing.

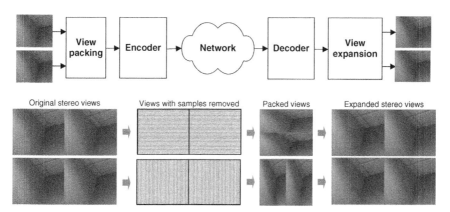

Figure 7.2   Sample removal, view packing, and view expansion (upconversion).

step to support spatial multiplexing. The packing method must be structured so that the views can be unpacked and upconverted (expanded) easily without unnecessary bit overheads. Pay TV service providers typically employ side-by-side (SbS) and top-and-bottom (TaB) view packing technologies to deliver S3D videos to the home. These frame-compatible technologies are designed to consume the same bandwidth as 2D transmission by subsampling the horizontal or vertical resolution of each image by half to fit into a single video frame.

7.1.3   View Expansion and Display

By splitting each frame into two images, one for each eye, the horizontal or vertical spatial resolution is reduced to form a SbS image or as a TaB image. The 3D TV or set top box (STB) doubles the length of each image and then displays those images sequentially for the viewing glasses at the regular frame rate. This requires only a simple firmware update and no new STB hardware. While there may not be any visible artifacts, the overall spatial resolution and picture quality of the S3D video is reduced compared to temporal multiplexing. For instance, the resolution of a 1080p video (about 2 million pixels) is immediately reduced to a level similar to 720p (about 1 million pixels) with view packing.

7.1.4   View Packing Methods

In SbS multiplexing, alternate columns of samples in each view are removed and the remaining samples are packed together. In TaB multiplexing, alternate rows of samples in each view are removed and the remaining samples packed together. As an alternative, alternate samples from each row of samples can be removed, giving rise to a checkerboard (also called quincunx) pattern, which can be row or column packed. A single iteration of sample removal is usually applied to pack two stereo views in a single frame. Two or more iterations are required to pack four or more views. Each iteration removes the same number of samples, effectively reducing the spatial resolution of each view by half. An odd iteration of row or column sample removal results in fewer adjacent samples for interpolation or concealment during upconversion. In contrast, checkerboard sample removal is unbiased for odd or even iterations as the loss in spatial resolution is equally distributed in the vertical and horizontal directions. However, this process may introduce higher distortion when the views are unpacked. Note that removing samples column-wise and row-wise using two iterations is also unbiased.

7.2 MULTIVIEW CODING

Multiview coding (MVC) utilizes prediction between different views to exploit the redundancy and improve the coding efficiency. H.264 includes an MVC extension that provides new techniques for reduced decoding complexity and scalable multi-view operations, including marking of reference frames and efficient view switching, such as using different sequence parameter sets (SPSs). Similarly, HEVC also provides a MVC extension [1] and caters for UHD 3D videos. To enable firmware upgrades, multiview HEVC only contains high-level syntax changes compared to HEVC and no changes to block-level processes. Any block-level process that is useful for multiview HEVC can only be enabled using hooks. Motion prediction hooks do not significantly impact single-view HEVC coding because they are designed to improve inter-view coding. As 2D TVs are still prevalent, the MVC standard achieves backward compatibility by defining the bitstream so that a compliant 2D decoder can decode a single 2D view and discard the rest of the data whereas a compliant MVC decoder can decode all views and generate the 3D video. MVC utilizes efficient ways to buffer the frames used for prediction, and enable parallel processing of separate views.

7.2.1 MVC Bitstream

A MVC bitstream comprises a base view and one or more nonbase (secondary) views. To improve coding efficiency, a nonbase view can utilize other views for inter-view prediction using temporal motion tools such as disparity vectors. In addition, the parallel decoding supplementary enhancement information (SEI) provides coded views with systematic constraints so that an macroblock (MB) in a specific view depends only on a subset of decoded MBs in other views. While MVC does not add new coding tools (e.g., new MB modes) on top of codecs such as H.264 or H.265, MVC does specify new high-level syntax for the network abstraction layer (NAL) unit and slice headers. For example, a new NAL unit type called coded MVC slice is used for coding nonbase views. It consists of a new 4-byte header that includes the priority ID, temporal ID, anchor frame flag, and inter-view flag. Anchor frames can be decoded without previous frames and thus serve as random access points. Random access at non intracoded frames is also possible using gradual decoding refresh (GDR). Anchor and nonanchor frames can have different dependencies, which can be signaled in the SPS.

7.2.2 2D to 3D Conversion

The efficient conversion of existing 2D videos to 3D has become a major component of 3D content production and delivery. One way to convert 2D to 3D is to generate a depth map or view dependency tree at the decoder. The received MV can be used to derive the displacement of objects over two consecutive frames. In doing so, the depth information need not be sent, thus conserving bandwidth. Alternatively, 2D plus depth or 2D plus difference methods can be applied. 2D plus depth has been developed to support multiview and freeview displays, not stereoscopic displays. With freeviewing, both left and right images are always visible. The 3D glasses allow the left eye to see image intended for left eye. In 2D plus difference, the main view is the left view while difference is the left view minus the right view. This difference should contain less information than the right view. The left and right views can be temporally multiplexed into a single MVC bitstream by alternating frames or fields in a left-right-left-right sequential pattern, which preserves the full spatial resolution at the expense of compromising the temporal resolution.

7.2.3 H.264 Multiview Coding Extension

The JM reference software contains open source code that implements the H.264 MVC extension. The software supports two or more views for S3D and autostereoscopic 3D displays. More information on MVC video streaming can be found in [2]. Although full-resolution views are more common, packed views can also be used. Views are input as separate videos into the encoder, which produces a single file with temporally multiplexed MVC-encoded views. These views are output as separate videos at the decoder. Several variable bit rate (VBR) HD videos encoded in H.264 are tested and listed in Tables 7.1 and 7.2.

7.2.4 MVC Inter-view Prediction

The basic concept of inter-view prediction is to exploit temporal motion tools such as disparity vectors. This is enabled by the *MVCEnableInterView* option in the JM

Table 7.1 Sample VBR H.264 Videos.

Name	HD Resolution	Frame Rate (Hz)	Duration (s)	Average Bit Rate (Mbit/s)
Magicforest	$1920 \times 1080p$	25	636	7.717
Magicforest Left View	$1440 \times 1080p$	25	636	6.467
Magicforest Right View	$1440 \times 1080p$	25	636	5.042
Avatar	$1280 \times 720p$	30	208	1.257
Forest	$1280 \times 720p$	25	253	3.020
Coke	$1280 \times 720p$	24	121	2.376
Flowers	$960 \times 720p$	30	384	0.759
Fountain	$1280 \times 720p$	30	67	6.750

Table 7.2   Sample VBR H.264 Videos.

Name	HD Resolution	Aspect Ratio	2D or S3D	Content Complexity
Magicforest	1920×1080p	16:9	SbS	Moderate activity. High scene complexity
Magicforest Left View	1440×1080p	4:3	2D	Moderate activity. High scene complexity. Original base view of *Magicforest*
Magicforest Right View	1440×1080p	4:3	2D	Moderate activity. High scene complexity. Original nonbase view of *Magicforest*
Avatar	1280×720p	16:9	SbS	High activity. Moderate scene complexity
Forest	1280×720p	16:9	SbS	Moderate activity. High scene complexity
Coke	1280×720p	16:9	SbS	High activity. Moderate scene complexity
Flowers	960×720p	4:3	TaB	Low activity. Low scene complexity
Fountain	1280×720p	16:9	2D	Low activity. High scene complexity

software. Figure 7.3(a) shows the simulcast H.264 coding of the left and right views of a S3D video using the IPB group of pictures (GOP) structure. Figure 7.3(b) shows the MVC coding of the left and right views of the same video. For simulcast H.264 coding, temporal redundancy (between frames) and intraview spatial redundancy (within a frame) are exploited to achieve coding gain. However, for MVC coding, in addition to temporal redundancy and intraview spatial redundancy, inter-view temporal redundancy between the views within the same frame is exploited to achieve higher coding efficiency. In order to reduce complexity, only the nonbase view is inter-view predicted using the base view. The base view is encoded in a similar way to single-view H.264 encoding. This also helps existing H.264 decoders (not supporting MVC) to decode single views from an MVC-encoded 3D video, thereby maintaining backward compatibility.

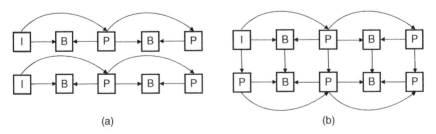

(a)   (b)

Figure 7.3   (a) Simulcast H.264 coding and (b) MVC coding.

Since the prediction is adaptive, the best predictor among the temporal and inter-view references can be selected on a MB basis. MVC makes use of the flexible reference frame management capabilities that are already available in H.264 by making decoded frames from other views available as reference frames for inter-view prediction. Specifically, reference frame lists are maintained for each frame to be decoded in a given view. Each list is initialized as usual for the single-view video, which may include temporal reference frames used to predict the current frame. Additionally, inter-view reference frames are included in the list and made available for prediction of the current picture. MVC does not allow prediction of a frame in one view at a given time using a frame from another view at a different time.

7.2.5 MVC Inter-view Reordering

Just as it is possible for an encoder to reorder the positions of the reference frames in a reference frame list (including temporal reference frames), MVC can also place inter-view reference frames at any desired position in the list. The core MB level and lower level decoding modules of the MVC decoder are the same, regardless of whether a reference frame is a temporal or inter-view reference.

7.2.6 MVC Profiles

Currently, there are two profiles defined by MVC: multiview high profile and the stereo high profile. Both profiles are based on the high profile of H.264 with the following differences:

- Multiview high profile supports multiple views and does not support interlace coding tools;
- Stereo high profile (supported by Blu-ray) is limited to two views and supports interlace coding tools.

7.2.7 Comparing MVC with 2D H.264 Video Coding

MVC requires higher encoding/decoding complexity than 2D H.264 video coding because multiple views from different videos are encoded/decoded simultaneously in order to exploit spatial redundancy. However, as shown in Table 7.3, the improvement in coding efficiency for MVC is marginal when compared to H.264 (less than 1% for two-view S3D). The video quality is hardly affected. In addition to the higher bit overheads associated with MVC coding (due to the need for inter-view prediction), the key reasons for this phenomenon are discussed in the next section. Note that the combined encoded file sizes for the left and right views are similar for both MVC and H.264. Table 7.4 shows the improvements in coding efficiency as the number of views per frame increases or equivalently, as the video resolution for each view decreases (3.5% for 4 views). In addition, the H.264 file size for 4-packed video is roughly the same as a 2-packed video if the video resolution is the same.

Table 7.3   MVC Versus H.264 for *Magicforest Left View* and *Magicforest Right View*
(100 frames, QP = 28, IPB GOP).

	MVC High Profile*		H.264 High Profile	
	Left View	Right View	Left View	Right View
Average Y-PSNR (dB)	35.72	37.10	35.72	37.06
Average U-PSNR (dB)	39.94	40.30	39.95	40.27
Average V-PSNR (dB)	42.87	43.52	42.87	43.53
I-coded bits	1,324,344	0	1,324,280	1,008,848
P-coded bits	10,640,320	7,327,864	10,649,728	6,532,576
B-coded bits	1,797,944	889,616	1,800,760	878,384
Total bits (1 view)	13,762,800	8,217,656	13,774,960	8,420,000
Total bits (2 views)	21,980,456		22,194,960	

*Multiview high profile and stereo high profile give the same results.

Table 7.4   2-Packed*MVC Versus 2-Packed† and 4-Packed‡ H.264 for *Coke* and *Forest*
(120 frames, QP = 28, IPB GOP).

	MVC High Profile		H.264 High Profile		
	2-Packed Left Views	2-Packed Right Views	2-Packed Left Views	2-Packed Right Views	4-Packed Left and Right Views
Average Y-PSNR (dB)	37.25	36.84	38.98	38.87	37.20
Average U-PSNR (dB)	44.09	43.94	44.89	44.80	44.05
Average V-PSNR (dB)	46.32	46.15	46.48	46.45	46.25
I-coded bits	384,296	0	561,072	596,792	795,208
P-coded bits	7,836,872	8,648,240	11,406,552	12,161,496	16,195,624
B-coded bits	1,033,632	567,216	4,538,224	4,857,096	2,141,432
Total bits (2 views)	9,254,976	9,215,624	16,506,024	17,615,560	
Total bits (4 views)	18,470,600				19,132,440

*Comprises two views per video frame, 640 × 720 overall resolution per frame.
†Comprises two views per video frame, 1280 × 720 overall resolution per frame.
‡Comprises four views per video frame, 1280 × 720 overall resolution per frame.

7.3   CORRELATION BETWEEN LEFT AND RIGHT VIEWS IN S3D VIDEOS

Figure 7.4(a) shows that the correlation between two SbS views for *Avatar* and *Magicforest* (first 200 frames) is weak with less than 20% giving a Y-PSNR of 40 dB or more. The right (nonbase) view is then horizontally shifted one column of samples at a time relative to a static left (base) view. The Y-PSNR of the right view is computed using the left view as a reference. This is repeated for up to 10 columns to the left and right of the static view. The maximum Y-PSNR improvement relative to the case

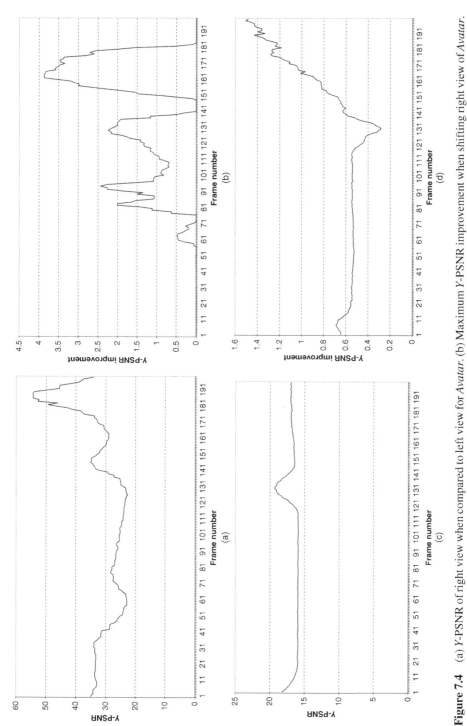

Figure 7.4 (a) *Y*-PSNR of right view when compared to left view for *Avatar*. (b) Maximum *Y*-PSNR improvement when shifting right view of *Avatar*. (c) *Y*-PSNR of *Magicforest Right View* when compared to *Magicforest Left View*. (d) Maximum *Y*-PSNR improvement when shifting *Magicforest Right View*.

229

when there is no shift (i.e., Figure 7.4(a)) is selected among the 20 Y-PSNR values. The results are shown in Figure 7.4(b). Less than 10% of the frames for *Avatar* show an improvement of over 3 dB. For the higher resolution *Magicforest Left View* and *Magicforest Right View*, the correlation is the weakest (Figure 7.4(c) and (d)). These results suggest that inter-view sample interpolation methods may lead to degradation in video quality, especially for higher resolution videos. In addition, these methods create dependency between the views and require more processing time compared to intraview methods. Thus, MVC may not provide significant coding efficiency when compared to H.264. The process of view shifting is repeated in Figure 7.5 except that this is conducted for successive frames, as opposed to left and right views. As can be seen, there is more correlation with the current frame when the previous frame is shifted. The drop in Y-PSNR in Figure 7.5(b) is due to a gradual scene change that is applied in a vertical fashion.

7.4 VIEW EXPANSION VIA SAMPLE INTERPOLATION

Sample interpolation is a concealment technique that is employed by many HDTV vendors to improve the visual quality and frame rate of fast action movies and sports, even when the original video is captured at a lower frame rate. For example, many 3D HDTVs are able to convert 1080p 24 Hz SbS Blu-ray output to 240 Hz (120 Hz for each view) using sample interpolation. Unlike frame repetition or sample copy, using sample interpolation to generate intermediate views or frames smoothes motion detail and enhances image clarity and sharpness. This can be done spatially (i.e., within a frame) or temporally (i.e., between frames). To illustrate spatial interpolation, we employ a block of 8×8 samples as shown in Figure 7.6. Spatial interpolation after one iteration of sample removal is straightforward and can be applied directly to SbS and TaB S3D videos – simply use the nearest available samples for interpolation. Spatial interpolation after two iterations of sample removal is somewhat challenging. If the same checkerboard pattern is successively applied to the original and packed views, a pattern identical to row–column removal (Figure 7.7(b)) results. However, if complementary checkerboard is used, a biased pattern results. Figure 7.7(a) shows a new checkerboard sample removal method where alternate columns of samples are shifted vertically by one sample. Figure 7.8 shows the results of concealing the samples. In Figure 7.8(a), the black cell uses the weighted average of the top and bottom neighbors for interpolating its value. The gray cell uses left, right-top, and right-bottom neighbors whereas the white cell uses left-top, left-bottom, and right neighbors. Likewise in Figure 7.8(b), the black cell uses left and right neighbors for interpolation. The gray cell uses top and bottom neighbors whereas the white cell uses left-top, left-bottom, right-top, and right-bottom neighbors.

7.4.1 Impact of Sample Interpolation

We compare the video quality after two iterations of sample removal in Tables 7.5 and 7.6. As can be seen, row–column sample removal consistently achieves better

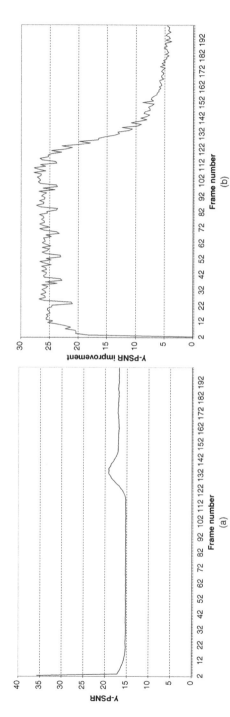

Figure 7.5 (a) *Y*-PSNR between current and previous frame of *Magicforest Left View.* (b) Maximum *Y*-PSNR improvement when shifting previous frame of *Magicforest Left View.*

1	2	3	4	5	6	7	8
9	10	11	12	13	14	15	16
17	18	19	20	21	22	23	24
25	26	27	28	29	30	31	32
33	34	35	36	37	38	39	40
41	42	43	44	45	46	47	48
49	50	51	52	53	54	55	56
57	58	59	60	61	62	63	64

Figure 7.6   Original block of 8×8 samples.

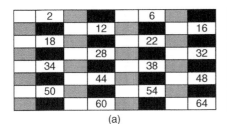

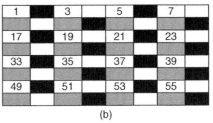

Figure 7.7   Remaining samples after two iterations of sample removal ($4\times$ compression): (a) checkerboard removal and (b) row–column removal.

Figure 7.8   Concealing samples after two iterations of sample removal: (a) checkerboard removal and (b) row–column removal.

visual quality than checkerboard sample removal (possibly due to more structured sample removal and interpolation) although the overall average Y-PSNR for both methods remains high. Checkerboard sample removal also results in less efficient H.264 coding. This can be confirmed by evaluating the H.264 file sizes for *Magic-forest Left View* after one iteration of row, column, checkerboard (column packed), and checkerboard (row packed) removal. The JM-encoded file sizes are 8,087,176, 8,223,248, 9,467,664, and 9,165,680 bits respectively. The experiment was repeated for *Fountain* using the H.264 encoder and the sizes are 15,467,192, 15,572,912, 17,098,240, and 17,085,912 bits respectively. Note that the row-packed video is most efficiently encoded, possibly because fewer rows are removed compared to

Table 7.5   Checkerboard Versus Row–Column Interpolation for *Flowers* (120 frames).

Average Y-PSNR	Original Top View Versus Interpolated Top View (Two Iterations of Checkerboard, 4× Compression)	Original Top View Versus Interpolated Top View (Row–Column Removal, 4× Compression)
Y (dB)	51.82 (46.64)	53.81 (46.64)
U (dB)	56.92 (54.41)	59.42 (54.41)
V (dB)	58.24 (57.71)	60.30 (57.71)

Table 7.6   Checkerboard Versus Row–Column Interpolation for *Coke*, *Forest*, and *Avatar* (120 frames Each).

Average Y-PSNR	Original Left View Versus Interpolated Left View (Two Iterations of Checkerboard, 4× Compression)	Original Left View Versus Interpolated Left View (Row–Column Removal, 4× Compression)
Coke		
Y (dB)	41.72 (35.89)	45.82 (35.89)
U (dB)	54.51 (53.25)	58.95 (53.25)
V (dB)	55.32 (55.32)	59.68 (55.32)
Forest		
Y (dB)	37.36 (32.13)	40.90 (32.13)
U (dB)	50.84 (46.43)	54.33 (46.43)
V (dB)	52.95 (51.34)	56.61 (51.34)
Avatar		
Y (dB)	43.25 (41.70)	43.97 (41.70)
U (dB)	55.08 (51.75)	56.15 (51.75)
V (dB)	55.21 (52.17)	56.42 (52.17)

the column-packed case. The Y-PSNR values in parentheses are obtained using column or row copy methods (i.e., no interpolation). These methods are currently employed when views are unpacked and expanded, and are clearly less superior than interpolation schemes. In this case, the Y-PSNR values are the same regardless of whether checkerboard or row–column sample removal is performed during packing. Table 7.7 illustrates the performance of checkerboard and row–column interpolation under multiple iterations of sample removal. As can be seen, up to eight times compression efficiency is achievable while maintaining a high average Y-PSNR. Again, row–column interpolation achieves better quality than checkerboard interpolation.

7.4.2   Inter-view Versus Intraview Sample Interpolation

Inter-view sample concealment methods may perform poorly when compared to intraview methods. To confirm this, we evaluate the performance of intraview

Table 7.7 Interpolation Performance for *Avatar* (120 frames).

Average *Y*-PSNR	Original Left View Versus Interpolated Left View (Row Removal, 2× Compression)	Original Left View Versus Interpolated Left View (Row–Column Removal, 4× Compression)	Original Left View Versus Interpolated Left View (Row–Column Removal Followed by Column Removal, 8× Compression)
Checkerboard			
Y (dB)	50.78	44.27	37.29
U (dB)	58.28	52.63	48.75
V (dB)	59.66	54.32	50.73
Row–Column			
Y (dB)	53.53	46.41	44.30
U (dB)	59.91	56.54	52.90
V (dB)	61.37	58.18	54.61

sample interpolation and inter-view sample copy for *Magicforest Left View* and *Magicforest Right View*. We subject the base and nonbase views to one iteration of complementary information removal (e.g., even row sample removal on left view and odd row sample removal on right view). For intraview interpolation, two vertical neighboring samples are selected. For inter-view sample copy, the missing sample from the base view is obtained from the corresponding sample in the nonbase view. A similar procedure is repeated for *Avatar* with one iteration of row removal. In this case, inter-view interpolation uses two vertical neighboring samples from the base view, two horizontal neighboring samples from the nonbase as well as the sample in the nonbase view corresponding to the position of the missing sample in the base view. As seen in Table 7.8, the video quality using inter-view sample interpolation is severely degraded, again demonstrating the lack of direct correlation between the views.

Table 7.8 Intraview Versus Inter-View Interpolation (100 frames).

Average *Y*-PSNR	Intraview Interpolation	Inter-view Interpolation
Magicforest Left View		
Y (dB)	31.00	19.12
U (dB)	44.87	31.93
V (dB)	49.79	39.98
Avatar Base View		
Y (dB)	57.09	44.99
U (dB)	61.00	54.00
V (dB)	62.76	55.12

7.4.3   Interframe Versus Intraview Sample Interpolation

Figure 7.5(b) suggests that inter-frame sample interpolation may achieve better performance. Figure 7.9 shows the results obtained for the column-packed *Magicforest Left View* (one iteration). In Figure 7.9(a), intraview interpolation is performed using the left and right neighboring samples. In Figure 7.9(b), the corresponding sample from the previous frame is used to replace the missing sample. As can be seen, the video quality improves over intraview interpolation until a gradual scene change occurs. The correlation with the previous frame weakens as a result of this scene change, reducing the effectiveness of inter-frame sample copy. Figure 7.9(c) shows an integrated solution where the left and right samples, and the corresponding sample from the previous frame are averaged to predict the missing sample's value. In this case, the video quality becomes more consistent.

7.4.4   Impact of Quantization on Interpolated S3D Videos

We perform one iteration of column removal for *Magicforest Left View* (resulting in an SbS 3D video), encode and decode the video using H.264, and then apply spatial sample interpolation to the decoded video. As shown in Figure 7.10, employing less aggressive quantization (e.g., using a low QP value) will not improve the video quality of S3D views but increase the video file size instead. This is because the baseline video quality is degraded by sample removal in view packing. Losing half the samples in a view with each iteration of sample removal can have a detrimental impact on video quality. Table 7.9 shows that for a similar *Y*-PSNR, a higher QP value of 34 can be used for the original video, giving rise to an encoded file size that is two to four times smaller than the packed video. This is effectively four to eight times smaller since the number of samples in the packed video is half the original. Thus, coding efficiency of spatial multiplexing is significantly lower than temporal multiplexing of S3D views. This is because the spatial multiplexing requires subsampling of the original video frame, which is more disruptive to the encoder than temporal multiplexing. This can be compared to the use of rate capping employed by some video encoders (e.g., capped VBR, CBR), which lead to less efficient encoding or poorer video quality. Temporal sample interpolation based on motion vectors can be used to reduce the coded frame rate by half or less, further improving coding efficiency.

7.5   ANAGLYPH 3D GENERATION

Equation (7.1) is designed to generate optimized anaglyph 3D from the individual S3D views [3]. The vector $[r_a, g_a, b_a]$ represents the red, green, and blue (RGB) color components of the optimized anaglyph. The vectors $[r_1, g_1, b_1]$ and $[r_2, g_2, b_2]$ represent RGB color components of the left and right views respectively. For optimized anaglyph, there is partial color reproduction and almost no retinal rivalry (which is caused by brightness differences of the colored objects). The interpolated TaB and anaglyph views for *Flowers* after two iterations of interpolation are

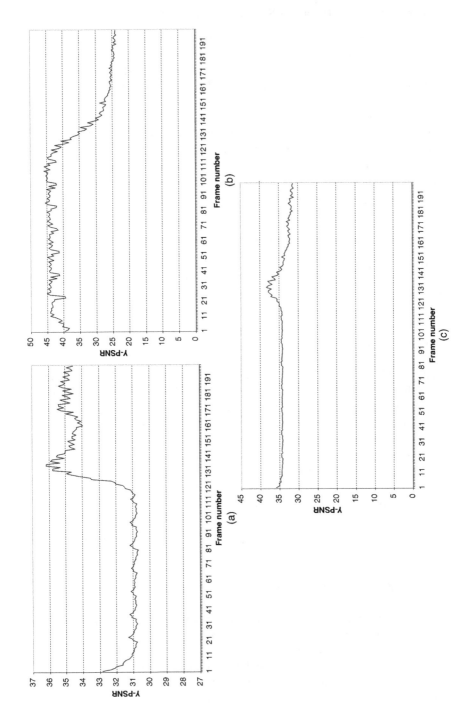

Figure 7.9 Concealing samples after one iteration of column removal: (a) intraview sample interpolation, (b) inter-frame sample copy, and (c) inter-frame sample interpolation.

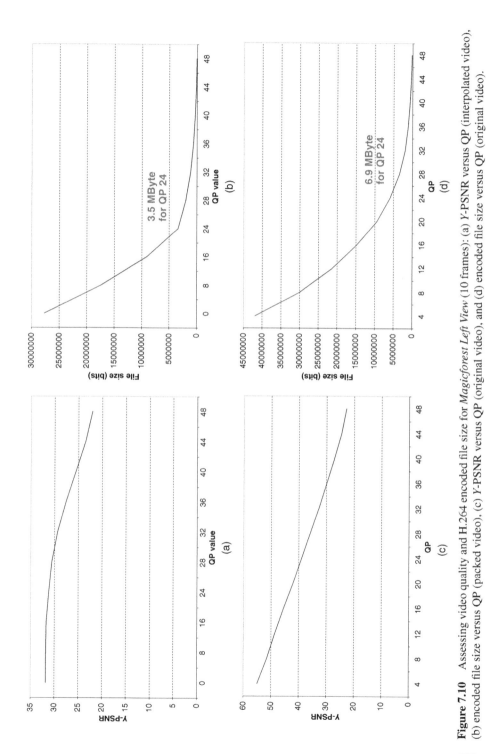

Figure 7.10 Assessing video quality and H.264 encoded file size for *Magicforest Left View* (10 frames): (a) *Y*-PSNR versus QP (interpolated video), (b) encoded file size versus QP (packed video), (c) *Y*-PSNR versus QP (original video), and (d) encoded file size versus QP (original video).

Table 7.9   Impact of Quantization on Original and Column-Packed *Magicforest*
Left View.

Video	Resolution	Quantization Parameter (QP)	Encoded File Size (bits)	Decoded Y-PSNR (dB)	Decoded U-PSNR (dB)	Decoded V-PSNR (dB)
10 frames						
Original	$1440 \times 1080p$	34	1,577,256	31.70	40.40	43.79
Column removed	$720 \times 1080p$	24	3,403,104	31.19*	40.49*	43.81*
100 frames						
Original	$1440 \times 1080p$	34	4,121,064	31.20	37.62	41.08
Column removed	$720 \times 1080p$	24	16,912,128	30.47*	40.08*	43.10*

*Includes spatial sample interpolation.

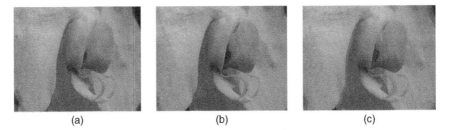

(a)   (b)   (c)

Figure 7.11   Interpolated snapshots for column-packed *Flowers* (4× compression): (a) top
view, (b) bottom view, and (c) anaglyph.

shown in Figure 7.11:

$$
\begin{bmatrix} r_a \\ g_a \\ b_a \end{bmatrix} = \begin{bmatrix} 0 & 0.7 & 0.3 \\ 0 & 0 & 0 \\ 0 & 0 & 0 \end{bmatrix} \begin{bmatrix} r_1 \\ g_1 \\ b_1 \end{bmatrix} + \begin{bmatrix} 0 & 0 & 0 \\ 0 & 1 & 0 \\ 0 & 0 & 1 \end{bmatrix} \begin{bmatrix} r_2 \\ g_2 \\ b_2 \end{bmatrix} \tag{7.1}
$$

7.5.1   H.264 Coding Efficiency for Anaglyph Videos

The H.264 coding efficiency for anaglyph videos is shown in Table 7.10. We employ
an encoder with a QP of 28. The average size is computed as (left view size + right
view size)/2. The relative overhead is then computed as (anaglyph size − average
size)/(average size) × 100%. Interpolation is first performed on the raw YUV video
files of *Forest*, *Coke*, and *Flowers* to convert the packed views to the original resolution. The YUV file is then converted to RGB format for anaglyph generation, which is
converted back to YUV for H.264 coding. As can be seen, there is a 20–25% increase
in overheads if the interpolated views are used. However, the overheads are negative

Table 7.10   Optimized Anaglyph H.264 Coding Efficiency.

	Left View (bytes)	Right View (bytes)	Optimized Anaglyph (bytes)	Overhead (%)
Forest (120 frames)	2,094,977	2,244,695	2,652,667	22.25
Coke (120 frames)	844,876	879,426	1,042,503	20.92
Flowers (120 frames)	199,663	193,719	247,114	25.64
Magicforest Left/Right Views (350 frames)	13,718,082	10,032,586	11,025,961	−7.15

if the original views of *Magicforest* are used. Note that in this case, the H.264 encoded anaglyph file size can be predicted since it is roughly equal to the sum of 1/3 left view file size and 2/3 right view file size, consistent with equation (7.1).

7.5.2   Delta Analysis

The use of a delta value to represent the difference between the original sample value and value obtained using interpolation may enhance the visual quality of the reconstructed video. In theory, the delta values can range from −255 to +255 depending on the value of the original sample and value obtained using interpolation. In order to represent the range of values from −255 to +255, 9 bits are required, which implies more overheads than sending the actual sample value! In practice, fewer bits are required due to correlation between adjoining samples. For example, Figure 7.12 shows the delta occurrence results for the Y, U, and V components of the base view of *Magicforest*. Eight bits are required to represent the exact sample values of the Y component (delta values range from −66 to +74) but only five bits are required for the U and V components (delta values range from −11 to +14 and from −9 to +10 respectively). Figure 7.13 shows the results for *Fountain*. It can be observed from these two videos that the probability distribution for the Y values is broader than the variation for U and V values due to more aggressive subsampling for the U and V components (we employ the 4:2:0 color format). Following this observation, the set-based delta approach is applied to the Y component to improve the Y-PSNR. Four bits are allocated to represent the delta to reduce the overheads using the following approximations:

- *Set 0 Delta:* Use 4 bits to represent the least significant 4 bits (bits 0–3) of the delta value. Only delta values in the range −8 to +7 can be represented using this approach. Delta values less than −8 and greater than +7 are upper-bounded to −8 and +7 respectively.
- *Set 1 Delta:* Use 4 bits to represent bits 4 through 1 of the delta value. Even numbers in the range −16 to +15 can be represented using this approach. If the original delta value is an odd number, the value it finally takes is equal to the quotient obtained by performing an integer division of delta with 2.

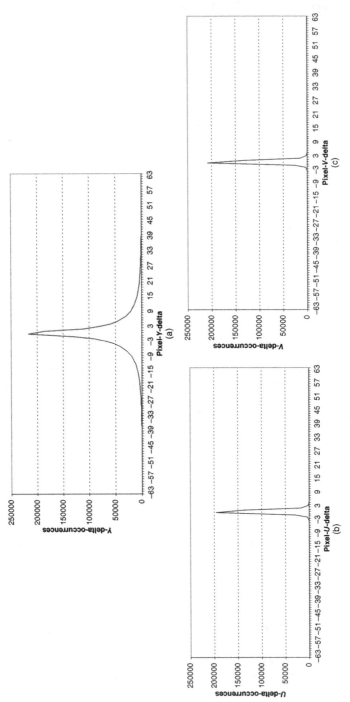

Figure 7.12 (a) *Y*-, (b) *U*-, and (c) *V*-delta occurrences for *Magicforest* (100 frames).

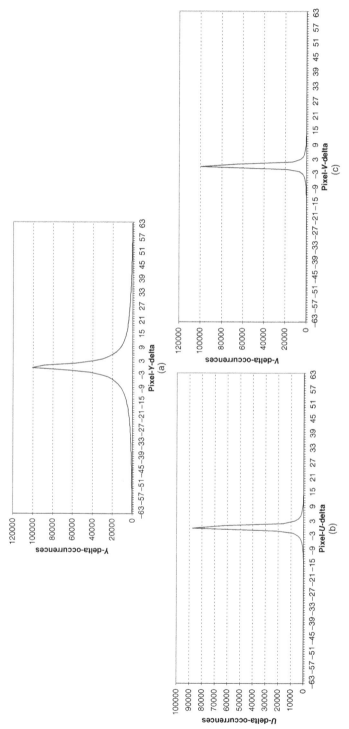

Figure 7.13 (a) Y-, (b) U-, and (c) V-delta occurrences for *Fountain*.

Table 7.11   Delta Performance after One Iteration of Checkerboard Sample Removal.

	Average Y-PSNR for *Magicforest* (dB)	Average Y-PSNR for *Fountain* (dB)
Original versus interpolated	35.55	25.24
Original versus interpolated with Set 0 Delta	40.36	26.94
Original versus interpolated with Set 1 Delta	42.73	28.30
Original versus interpolated with Set 2 Delta	44.06	31.06
Original versus interpolated with Set 3 Delta	41.55	35.21
Original versus interpolated with Set 4 Delta	38.44	34.83

- *Set 2 Delta:* Use 4 bits to represent bits 5 through 2 of the delta value. Multiples of 4 in the range −32 to +31 can be represented using this approach. If the original delta value is not a multiple of 4, the value it finally takes is equal to the quotient obtained by performing an integer division of delta with 4.

The results are shown in Table 7.11. For *Magicforest*, the average Y-PSNR monotonically decreases from Set 0 to Set 2. However, for *Fountain*, Set 2 yields the maximum Y-PSNR. This phenomenon can be explained as follows. For *Magicforest*, samples with delta values ranging from −16 to +15 represent 93% of the total samples. For *Fountain*, samples with delta values ranging from −16 to +15 represent only 80% of the total samples.

7.5.3   Disparity Vector Generation

Disparity estimation is useful for recovering the depth map of a pair of stereo views and can potentially improve the video quality of S3D videos. Quality metrics may include depth range, vertical misalignment, and temporal consistency. The distinction between motion and disparity vectors is shown in Figure 7.14. We employ the disparity vector to recreate the nonbase view from the base view. This may be used to further reduce the bandwidth requirements or improve the video quality of the base view. Two views of an S3D video are passed to a disparity vector generation algorithm that partitions each frame of the nonbase view into $N \times N$ blocks of samples. For each sample block in the nonbase view, the algorithm attempts to find the best match based on the sum of absolute differences (SAD) with blocks of the base view in the same row (i.e., vertical parallax between the two views is assumed to be zero).

The block under investigation in the nonbase view is scanned against a total of 256 blocks from the base view (ranging from a left shift of 128 samples to a right shift of 127 samples). The sample shift value (ranging from 0 to 255) corresponding to the best block match is recorded as the disparity vector. For a block size of $N \times N$, (Width/N) × (Height/N) disparity vectors are computed for each frame. This translates to $1/N^2$ reduction in the overheads required to represent disparity vectors when compared to the data required to represent the raw samples. The disparity data generated as described earlier can be efficiently coded using platelet-based coding

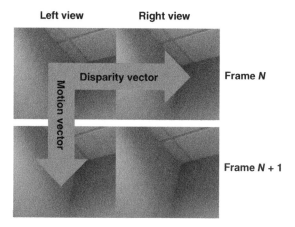

Figure 7.14   Distinction between motion and disparity vectors.

Table 7.12   Average *Y*-PSNR Using Disparity Vectors (15 frames).

	N	Average *Y*-PSNR for *Avatar* (dB)	Average *Y*-PSNR for *Magicforest* (dB)
Left view versus right view (no disparity vectors)	NA	30.06	16.94
Right view versus right view generated	4	30.40	27.49
Right view versus right view generated	2	33.48	32.47
Right view versus right view generated	1	36.69	37.42

methods [4]. MVC coding methods are not recommended for coding disparity data as they may cause coding errors leading to visible distortions in the synthesized view.

The generation of the right (nonbase) view from the left (base) view using disparity vectors is straightforward. For each block in the right view, the corresponding disparity vector is used to obtain the block from left view that minimizes the difference in sample values. As seen from Table 7.12, the improvement in *Y*-PSNR can be significant for smaller block sizes and this is confirmed with subjective evaluation.

REFERENCES

1. G. Sullivan et al., "Standardized Extensions of High Efficiency Video Coding," IEEE Journal of Selected Topics in Signal Processing, Vol. 7, No. 6, December 2013, pp. 1001–1016.
2. Y. Wang and T. Schierl, "RTP Payload Format for MVC Video," IETF Draft, June 2012, https://tools.ietf.org/html/draft-ietf-payload-rtp-mvc-02.

3. Optimized Anaglyph Generation, http://3dtv.at/Knowhow/AnaglyphComparison_en.aspx.

4. Y. Morvan, P. H. N. de With, and D. Farin. "Platelet-Based Coding of Depth Maps for the Transmission of Multiview Images," *Proceedings of SPIE, Stereoscopic Displays and Applications*, Vol. 6055, January 2006, pp. 93–100.

HOMEWORK PROBLEMS

7.1. Twin-packed views may lead to poorer coding efficiency when compared to single-packed views. From Table 7.4, can we conclude that 4-packed views are encoded more efficiently than twin-packed views? Note that the Y-PSNR for both cases are quite similar. By comparing the MVC efficiencies for Tables 7.3 and 7.4, can we conclude that MVC is more efficient as more views are packed in a frame?

7.2. Unlike 2D videos, the key subjects in 3D videos can be emphasized with depth information and background objects need not be blurred. Explain whether this implies that the overall video quality of 3D videos is better than 2D.

7.3. Some people claim that half-resolution SbS or TaB formatting is an interim solution for transporting 3D videos with an existing network infrastructure. Explain whether the existing infrastructure can support full-resolution temporal multiplexed 3D videos.

7.4. When packing S3D views using the row, column, and checkerboard methods, the same number of samples is removed in each iteration. Explain why row-packed views lead to the best Y-PSNR when the views are expanded.

7.5. Compute the display aspect ratio for a 720p 3D movie. Note that there are two answers to this question.

7.6. There are two general ways to verify the characteristics of a coded video. First, the frame sizes can be extracted and then summed up to give the overall video file size. Second, the natural video frame rate can be used to derive the frame duration. When multiplied by the total number of frames, this should match the duration of the video. Can the same methods be applied to 3D videos?

7.7. Although the video bit rate a key metric for assessing digital video quality, as illustrated in Section 7.4.4, the subsampled 3D video bit rate can be easily increased with virtually no improvement in the video quality. Thus, a high bit rate 3D video will actually have poor visual quality if it is upconverted from a subsampled version. How can the original (unsampled) coded bit rates of such videos be estimated?

8

VIDEO DISTRIBUTION AND STREAMING

Adaptive bit rate or adaptive streaming is the most efficient and convenient way to transmit video entertainment to many users. Varying network conditions and computing resources of the user device are taken into account before the appropriate quality profile is requested by the device and the video is sent by the server. Devices with high-speed connectivity may receive high-quality streams while others with lower speeds may receive lower quality streams. Thus, users of multiscreen devices may enjoy an uninterrupted and stutter-free streaming experience without service providers catering to the lowest common denominator in the quality level or display resolution. Adaptive video streaming is superior to delivering a video at a single bit rate because the quality of the video can be adjusted midway to be as good or bad as the client's available network speed instead of buffering or interrupting playback that can happen when a client's network speed cannot support the quality of video. Adaptive streaming has been adopted on a wide scale and has been demonstrated by ESPN3 to be robust enough to broadcast over 10 million simultaneous live streams at the 2010 World Cup. In this chapter, we will evaluate important metrics (e.g., video chunk duration and bit rate) that impact the efficiency of adaptive streaming. An in-depth review on the practical performance of popular adaptive streaming platforms (Apple's Hyper-Text Transfer Protocol Live Streaming (HLS) and Microsoft's smooth streaming (MSS)) over a variety of wireless networks will be presented.

Next-Generation Video Coding and Streaming, First Edition. Benny Bing.
© 2015 John Wiley & Sons, Inc. Published 2015 by John Wiley & Sons, Inc.

8.1   ADAPTIVE VIDEO STREAMING

Adaptive streaming is a combination of scalable-quality video streaming and progressive download. It is a significant enhancement of progressive download, which enables the playback of a video when a specified amount of data becomes available prior to completing the full download. Instead of adapting the public Internet to real-time multimedia traffic with new network protocols, adaptive streaming adapts media delivery to the Internet using conventional HTTP. Instead of maximizing throughput, adaptive streaming sends compressed video traffic while minimizing loss without necessarily increasing delay. This is in contrast to data buffering, which minimizes packet losses at the possible expense of increased delay. The video is fragmented into small chunks (or "streamlets") and then encoded with varying levels of video quality, each level requiring a specific bit rate (Figure 8.1). A higher quality level will require a higher bit rate. The chunks are hosted on a regular HTTP Web server and referenced using URLs. A client or user device requests the chunks sequentially from the server and downloads them for playback.

The video quality of the chunks (i.e., bit rate or resolution) can be switched dynamically and seamlessly to adapt to fluctuating link conditions and end-user device resources. In doing so, annoying video artifacts are minimized while smooth playback at the video's natural frame rate is maintained. For example, when bandwidth resources become scarce or the network becomes lossy, the client may request the server to deliver a chunk encoded with a lower quality. Users no longer need to manually select a bit rate or resolution for video playback as the device can now perform that function dynamically and automatically. Users may also enjoy fast start-up and seek times as playback control can be initiated using the lowest bit rate chunks and subsequently increased to higher rates.

On a managed network, since packet losses are not common, bit rate variation can be minimized by appropriately changing the quality of the chunks (Figure 8.2). For example, the quality of the chunks can be adapted to different content complexity. This will enable more efficient video delivery over fixed bandwidth channels in the network. Longer chunks can also be used, as opposed to the 2–10s chunks typically used for operations over unmanaged networks such as the public Internet. Adaptive streaming is a key technology enabler for efficient and cost-effective multiscreen video delivery to the TV, laptop, tablet, and smartphone as TV programs and the Internet converge over a variety of managed and unmanaged IP networks.

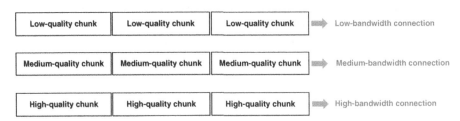

Figure 8.1   Quality (bitstream) switching in adaptive streaming.

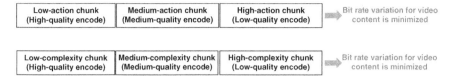

Figure 8.2   Minimizing bit rate variation with different quality chunks.

8.1.1   Playlists and Bandwidth Estimation

Many adaptive systems require a custom video player at the receiver to estimate the bandwidth, play the chunks, and request the appropriate chunks from the Web server when a switch to a different quality bitstream is desired. Since some or all of the chunks are pre-encoded in advance, the average rate of each quality profile can be precomputed by the server and relayed to the client in the form of a playlist or manifest file. For on-demand services, the playlist contains information on all chunks in the video. For live streaming, references to a sliding window of currently available chunks are provided in the playlist, which must be updated periodically. The client then uses these references to download chunks and play them back in sequence.

Because Web servers usually send data as quickly as possible using the available network bandwidth, the client can easily estimate the bandwidth using the chunk download times. The client's bandwidth calculation is normally repeated after every chunk download so that it can adapt to changing network bandwidth every few seconds. The client may also take into account its current memory resources, computing load, and rendered frame rates before requesting the appropriate quality chunks ahead of time. The server will in turn send the video chunk pre-encoded at a rate suitable for the connection. The bit rate scalability of adaptive streaming is particularly effective in minimizing the impact of the bit rate variability in wireless networks and the public Internet.

8.1.2   Quality (Bitstream) Switching

The key to the successful adaptive streaming is the ability to switch between different quality levels without a perceptible change in the video quality. In general, this requires a change in the QP values, which can apply to both variable bit rate (VBR) and constant bit rate (CBR)-coded videos. A high QP value is employed for more lossy encoding, which reduces the video quality and bit rate. For CBR-coded chunks, the QP values can be modified within each chunk in order to achieve the desired bit rate. For VBR-coded chunks, the change in the QP value can be achieved on a chunk-by-chunk basis. In this case, the overall bit rate variation can be minimized by employing different QP values for chunks that contain different types of content. For example, low action content that lead to low bit rates can be transmitted at higher quality (i.e., using chunks encoded with a low QP value). In general, a QP value that ranges between 16 and 24 may not lead to perceptible differences in quality although the difference in the rates can be substantial. However, a chunk encoded with a very high QP value of 40 can lead to perceptible video quality degradation.

8.2 VIDEO QUALITY AND CHUNK EFFICIENCY

Some adaptive streaming systems are "key frame aligned," which means an intra-coded frame is inserted at the start of each chunk to allow independent decoding and seamless switching between past and future chunks. Thus, closed GOPs with fixed lengths are employed and each chunk is sliced along GOP boundaries. An additional motivation for using aligned chunk boundaries is that they also serve as ad-insertion points, allowing ads to be inserted simply by chunk substitution, which is easier than splicing an unfragmented stream. However, this may reduce the compression efficiency because the video content at start of the chunk may not always require an intracoded frame. Depending on the content, one or more intracoded frames may be included within a single chunk. We analyze the compression efficiency and bit rates of the fragmented chunks that contain an intracoded frame at the start of the chunk. The SD and HD VBR videos are listed in Table 8.1.

8.2.1 Video Quality for Different VBR Chunk Durations

In Figures 8.3–8.5, the luminance peak-signal-to-noise ratio (Y-PSNR) for each frame is averaged over the entire VBR video. As can be seen, the chunk duration has a minor impact on the video quality. However, the QP value should be carefully calibrated to obtain an acceptable Y-PSNR. While subjective and objective video assessment methods exist, a simple and effective way to avoid motion and spatial artifacts is to restrict the QP value of each encoded frame to be below 30 in order to ensure a Y-PSNR of at least 40 dB. This applies to all encoding types: strict VBR, capped VBR, and CBR. The *Water* video shows a higher Y-PSNR variation because the content

Table 8.1 Fragmented VBR Videos.

Video	Resolution	Frame Rate (Hz)	Duration (s)	Average Bit Rate (Mbit/s)	Content Complexity
Water	1280 × 720p	30	12	2.463	Moderate activity with no scene change. Moderate to high scene complexity with bridge and water subjects sandwiched between blurred foreground and background
FCL	1280 × 720p	24	72	0.957	Moderate activity with 5 scene changes. High scene complexity
Office	1280 × 720p	30	20	9.711	No motion, moderate complexity
300	1280 × 720p	24	106	4.113	Moderate activity but high number of scene changes and includes audio

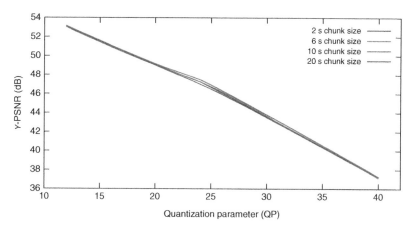

Figure 8.3 *Y*-PSNR for *FCL* with different chunk durations and QP values.

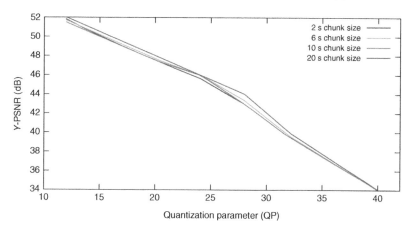

Figure 8.4 *Y*-PSNR for *Water* with different chunk durations and QP values.

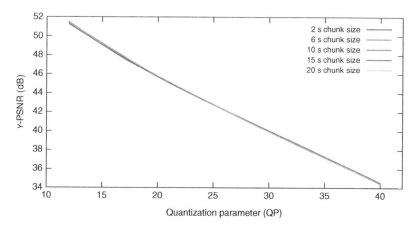

Figure 8.5 *Y*-PSNR for *Office* with different chunk durations and QP values.

is more challenging. For the *Office* video that contains no motion, the quantization process is very consistent for different chunk durations and almost linear for a broad range of QP values. In this case, a QP value of 28 leads to a *Y*-PSNR of 40 dB, which stays the virtually the same for all chunk durations.

8.2.2 VBR Chunk Bit Rate Versus Chunk Duration

The average chunk bit rate (bit/s) is the ratio of the chunk file size (bits) and the chunk duration (s). The chunk duration is the ratio of the number of video frames in the chunk and the frame rate. For example, a 10s chunk of a 30 Hz video contains 30×10 or 300 frames. Figures 8.6–8.8 show the H.264 VBR chunk bit rates for different QP values and chunk durations. As can be seen, shorter chunks generally lead to higher average chunk bit rates and this is primarily due to the intracoded frame at start of each chunk. The bit rate variability becomes lower for chunks encoded with high QP values (i.e., lower quality chunks). Content clearly plays a role since any chunk may contain high motion or complex video scenes, resulting in a higher average chunk bit rate. Because the *Office* video contains no motion, the dependency on content is removed. In this case, longer chunks are consistently coded more efficiently than shorter chunks. In addition, longer chunks lead to lower bit overheads resulting from fewer chunks. Note that the average chunk bit rate is almost constant for QP 24 and above.

8.2.3 VBR Chunk Efficiency Versus Chunk Duration

Instead of the chunk bit rate, we now examine the chunk efficiency in terms of the coded file size for different chunk durations. Depending on the duration of the entire video, the chunks may not always end at chunk boundaries. Tables 8.2 and 8.3 illustrate the H.264 coding efficiencies for VBR-coded chunks. The results are presented

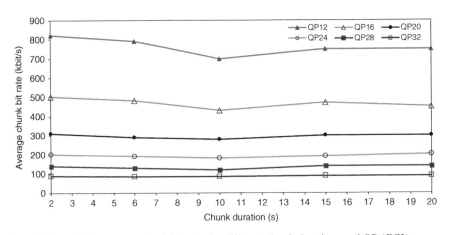

Figure 8.6 Average chunk bit rate for different chunk durations and QP (*FCL*).

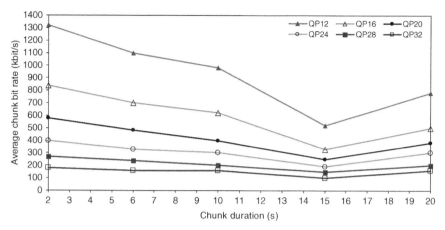

Figure 8.7 Average chunk bit rate for different chunk durations and QP (*Water*).

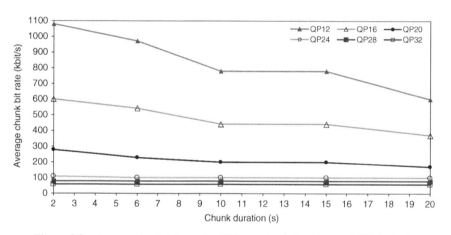

Figure 8.8 Average chunk bit rate for different chunk durations and QP (*Office*).

Table 8.2 H.264 VBR Coding Efficiencies for Different Chunk Durations (*Water*).

Quantization Parameter (QP)	Size for Six 2s Chunks (Mbyte)	Size for One 10s Chunk and One 2s Chunk (Mbyte)	Unfragmented Video (Mbyte)
20	9.138	8.241	6.335
28	4.275	4.020	3.649
40	0.960	0.917	0.912

Table 8.3   H.264 VBR Coding Efficiencies for Different Chunk Durations
(*Office*).

Quantization Parameter (QP)	Size for Ten 2s Chunks (Mbyte)	Size for Two 10s Chunks (Mbyte)	Unfragmented Video (Mbyte)
20	5.927	5.380	5.196
28	1.003	0.770	0.688
40	0.181	0.125	0.105

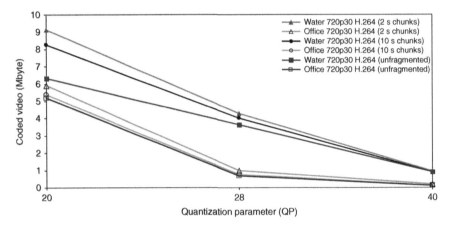

Figure 8.9   VBR coding efficiencies for fragmented and unfragmented videos.

graphically in Figure 8.9. Clearly, longer chunks lead to higher compression efficiencies (i.e., smaller coded file sizes) and this is independent of the content. Significant savings are achievable for videos with no motion or encoded with low QP values (i.e., at high-quality levels). These results confirm the observations of the previous section. The unfragmented videos incur the least coding overheads, thus leading to the best compression efficiencies.

8.2.4   Capped VBR Chunk Efficiency Versus Chunk Duration

In order to maintain a specific capped rate, Tables 8.4 and 8.5 show that the file sizes are almost identical for different chunk durations. The results are presented graphically in Figure 8.10. Thus, there is no clear trend in the coding efficiencies for different chunk durations. For a capped rate of 0.75 Mbit/s, the bit rate for the 2s fragmented video is $1.156 \times 8/12$ or 0.77 Mbit/s. Similarly, for a capped rate of 1.5 Mbit/s, the 2s bit rate becomes 1.55 Mbit/s and for a capped rate of 3.0 Mbit/s, the 2s bit rate becomes 3.1 Mbit/s. In general, there are more capping instances for longer chunks but this is content dependent.

Table 8.4   H.264 Capped VBR Coding Efficiencies for Different Chunk Durations (*Water*).

Capped VBR Rate (Mbit/s)	Size for Six 2s Chunks (Mbyte)	Size for One 10s Chunk and One 2s Chunk (Mbyte)	Unfragmented Video (Mbyte)
0.75	1.156	1.155	1.155
1.5	2.323	2.312	2.312
3.0	4.653	4.619	4.619

Table 8.5   H.264 Capped VBR Coding Efficiencies for Different Chunk Durations (*Office*).

Capped VBR Rate (Mbit/s)	Size for Ten 2s Chunks (Mbyte)	Size for Two 10s Chunks (Mbyte)	Unfragmented Video (Mbyte)
0.75	1.969	1.970	1.970
1.5	3.920	3.953	3.935
3.0	7.825	7.847	7.847

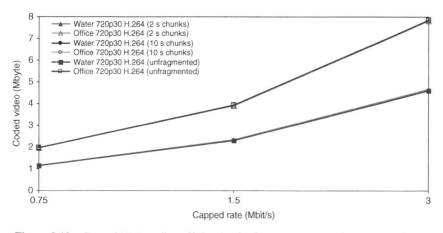

Figure 8.10   Capped VBR coding efficiencies for fragmented and unfragmented videos.

8.2.5   CBR Chunk Efficiency Versus Chunk Duration

The H.264 CBR file sizes in Tables 8.6 and 8.7 are all larger than the capped VBR file sizes. The results are presented graphically in Figure 8.11. Longer CBR chunks lead to slightly improved efficiency (<1.2%) and this trend is independent of the content. In general, CBR encoding requires more aggressive rate control when compared to

Table 8.6   H.264 CBR Coding Efficiencies for Different Chunk Durations (*Water*).

CBR Rate (Mbit/s)	Size for Six 2s Chunks (Mbyte)	Size for One 10s Chunk and One 2s Chunk (Mbyte)	Unfragmented Video (Mbyte)
0.75	1.197	1.183	1.183
1.5	2.390	2.366	2.366
3.0	4.791	4.737	4.737

Table 8.7   H.264 CBR Coding Efficiencies for Different Chunk Durations (*Office*).

CBR Rate (Mbit/s)	Size for Ten 2s Chunks (Mbyte)	Size for Two 10s Chunks (Mbyte)	Unfragmented Video (Mbyte)
0.75	2.074	2.029	2.029
1.5	4.065	4.003	4.003
3.0	8.027	7.952	7.952

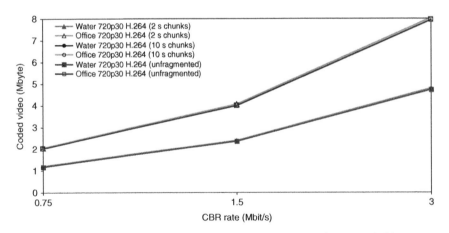

Figure 8.11   CBR coding efficiencies for fragmented and unfragmented videos.

capped VBR, which may lead to less efficient encoding or a compromise in the video quality.

8.2.6   Instantaneous and Average Rates for Different Chunk Durations

The instantaneous and average rates of the encoded chunks may differ significantly. These rates can be used to determine the complexity of the video content as well as

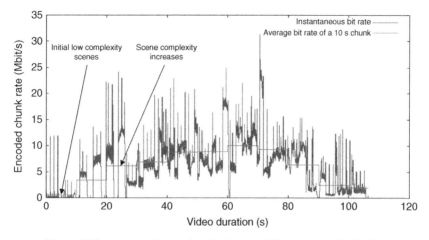

Figure 8.12   Average bit rate for 10s chunks (*300* encoded with QP 20).

the bandwidth requirements. The instantaneous rate is ratio of the size of the encoded frame (bits) in a chunk and the frame interval (s). For a 30 Hz video, the frame interval is 1/30s or 33.33 ms. The average chunk bit rate is the ratio of the chunk file size (bits) and the chunk duration (s). Figure 8.12 shows the rates for segmenting the VBR *300* H.264 video into 10s chunks using one quality level. The instantaneous bit rate can be three times greater than the average chunk bit rate. This difference reduces for shorter 2s chunks, as seen in Figure 8.13. A higher chunk bit rate variation is observed for 2s chunks compared to 10s chunks. Figure 8.14 illustrates the instantaneous rates for the unfragmented video, which incurs the least overheads. The 2s average rate for this video is also included in the figure. By comparing Figure 8.14 with Figures 8.12

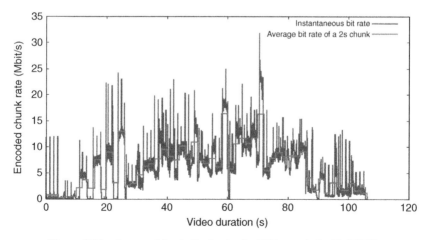

Figure 8.13   Average bit rate for 2s chunks (*300* encoded with QP 20).

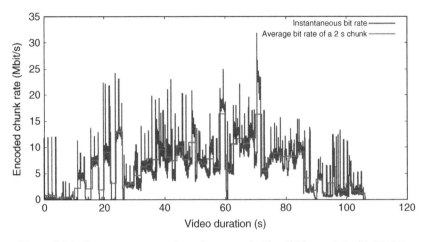

Figure 8.14   Instantaneous rate for unfragmented video (*300* encoded with QP 20).

and 8.13, it is quite evident that the instantaneous rates are closely matched, which implies the bit overheads for the one-level fragmented video are comparable to the unfragmented video.

Figures 8.15 and 8.16 show the instantaneous and average bit rates for the VBR *300* H.264 video when encoded with a high QP value of 40. Although the video quality and bit rate are lower, the characteristics remain similar to Figures 8.12 and 8.13 when a much lower QP value of 20 is used to encode the same video.

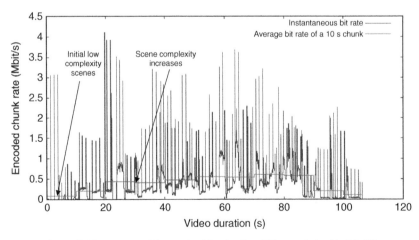

Figure 8.15   Average bit rate for 10s chunks (*300* encoded with QP 40).

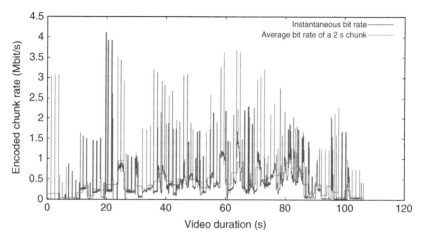

Figure 8.16   Average bit rate for 2s chunks (*300* encoded with QP 40).

8.3   APPLE HLS

HLS [1] supports both live broadcasts and video on demand (prerecorded content) for Apple iOS devices and PCs. It has become the adaptive streaming standard for Android and has expanded to over-the-top (OTT) streaming adapters (e.g., Chromecast, Roku) as well as smartphones and tablets. Thus, HLS is by far the most popular and mature adaptive streaming method, with many advanced features such as the ability to simultaneously enable multiple audio streams and audio formats, video profiles, languages, playlists, closed captions, and camera views. HLS employs MPEG-2 transport stream (TS) encapsulation for the compressed video and audio. An Apple application that delivers video longer than 10 min or greater than 5 Mbytes of data must use HLS and provide at least one stream at 64 kbit/s or lower rate. Multiple streams of the same media content are encoded at the server and the user device software can request the server to switch to different quality streams according to changes in network bandwidth or device resources.

8.3.1   Overview of HLS Operation

The source video and audio are first encoded into multiple bitstreams at different data rates and then segmented into a sequence of short chunks. The chunk duration specifies the time interval the source media is segmented into. Although the user may vary the chunk duration (typically from 2 to 10s), the standard duration for HLS streaming has been 10s. Apple recently reduced this duration to 9s. The chunk duration is the minimum amount of time a video plays at the current quality level before switching to the next level. Segmentation is required to enable seamless switching between levels.

The audio and video chunks should be precisely aligned across all levels. The segmented files are then placed on an HTTP Web server along with a text-based index file or playlist (with a .m3u8 extension) that lists the available streams.

The playlist is a text file based on Winamp's original m3u file format. The playlist can be refreshed periodically to support live broadcasts, where media chunks are generated continuously. The playlist provides the client with a URL of the segmented files. Each URL is published on the server and refers to a media chunk of a single video bitstream. The chunk can be specified as a byte range of a larger URL. This allows the chunks to be consolidated into larger files or a single large file, thereby reducing the number of files to manage and enabling proxy caching servers to look ahead when prefetching chunks.

A streaming client is made aware of the available streams with differing bit rates by the playlist. The encoding profile allows the player to identify and retrieve compatible streams. A link to the playlist can be embedded in a Web page or sent to an app. Apple recommends using the HTML5 video tag for deploying HLS on a website. The client reads the playlist and then requests the media chunks in order. Like the playlist, these media chunks are also referenced by URLs, typically by the names filesequence0.ts, filesequence1.ts, and so on. Once the chunks are downloaded, the player reassembles them so that the media can be presented to the user in a continuous stream. The player monitors the available bandwidth. If the bandwidth conditions mandate a stream change, the player checks the playlist for the location of alternative streams. Since HTTP is employed, an ordinary Web server with no streaming server software can be employed. Thus, all switching decisions can be made by the player.

All iOS 3.0 and later devices come with in-built HLS client software. The Safari 4.0 or later browser also supports HLS. HLS supports trick modes such as fast forward and reverse playback at specific time instants that are determined by the locations of the intracoded frames, which can be specified using the I frame only playlist. HLS supports CEA 608 and WebVTT closed captions. Nielsen ID3 broadcast tags can be translated to HLS content for mobile TV tracking. Nielsen is an audience-based TV ratings system. HLS also offers content protection via media encryption and user authentication over HTTPS, which are only available for prerecorded content. Currently, HLS supports AES-128 encryption using 32-byte keys. There are three ways in which encryption can be applied: using an existing key, using a randomly generated key, or using a new key that is generated for a set of video segments. Keys can be served over SSL for an added layer of encryption.

8.3.2 GOP Structure

Switching streams can be done flexibly because open GOP boundaries are permitted. Closed GOPs cannot contain any frame that references a frame in the previous or next GOP. In contrast, open GOPs may begin with one or more B frames that reference the last P frame of the previous GOP. This means that the start of each video chunk need not always start with an intracoded I frame or key frame (other than the first chunk of the video), allowing more efficiently encoded B frames to be packed into the GOPs, thereby improving overall compression efficiency. However, this may be done at the expense of more complex implementation of trick modes since the key frame within a chunk will have to be parsed and located, unlike the case when the key frame starts at every chunk.

A compatible MPEG-2 TS stream segmenter is provided by Apple. Alternatively, if a media file is already encoded using the supported codecs, a file segmenter can be used to encapsulate it in MPEG-2 TS and segment it into chunks of equal duration. The file segmenter contains a library of existing audio and video files for sending video on demand via HLS. The file segmenter performs the same tasks as the stream segmenter but takes files as input instead of streams.

Because HLS chunks may not start with intracoded frames, the client may download chunks from two different quality levels and switch the decoder between these levels on an intracoded frame that occurs within the chunk. However, this incurs additional bandwidth consumption as two chunks corresponding to the same portion of video are downloaded at the same time. Each TS chunk contains multiplexed audio and video information that cannot be transported separately, unless the audio and video bitstreams are segmented independently without any synchronization between them.

8.3.3 Super and Dynamic Playlists

HLS produces two types of playlist files for each quality level. The super (or master) playlist indicates the bit rates corresponding to the quality levels of the different streams, which are referenced using dynamic playlists. The dynamic playlist contains the current URLs of the most recent sequence of available video chunks. This playlist is refreshed periodically to keep it up-to-date. Thus, a new dynamic playlist is requested even when no change in quality is required. This increases the overall bandwidth consumption slightly since playlists are typically much smaller in size than the video chunks. The refresh duration can be specified and is subject to a minimum. The dynamic playlist can be downloaded every time a chunk is played.

Figures 8.17 and 8.18 show some examples of the super playlist and dynamic playlist files. Metadata carried in the playlist files are flagged as comments (i.e., lines preceded by #). The super playlist metadata includes the bit rate of the quality level, resolution, codec, and tags to identify different encodings of the same content. The dynamic playlist metadata contains the sequence number to associate chunks from different quality levels, chunk duration, information on whether chunks can be cached, location of decryption keys, stream type, and timing information. For the dynamic playlist file, the #EXT-X-MEDIA-SEQUENCE tag identifies the sequence number of the first chunk, 101.ts. It is used to align chunks from different quality levels. The #EXT-X-TARGETDURATION:2 tag indicates the expected duration of the chunks. The #EXT-X-KEY:METHOD=NONE tag shows that no encryption was used in this sequence of chunks. The #EXTINF:2 tags indicate the duration of each chunk. On-demand playlists are distinguished from live playlists by the #EXT-X-PLAYLIST-TYPE and #EXT-X-ENDLIST tags.

```
#EXTM3U
#EXT-X-STREAM-INF:PROGRAM-ID=1,BANDWIDTH=6370000,
http://example.com/playlist_QP20.m3u8
#EXT-X-STREAM-INF:PROGRAM-ID=1,BANDWIDTH=3901000,
http://example.com/playlist_QP24.m3u8
#EXT-X-STREAM-INF:PROGRAM-ID=1,BANDWIDTH=2274000,
http://example.com/playlist_QP28.m3u8
#EXT-X-STREAM-INF:PROGRAM-ID=1,BANDWIDTH=1241000,
http://example.com/playlist_QP32.m3u8
#EXT-X-STREAM-INF:PROGRAM-ID=1,BANDWIDTH=759000,
http://example.com/playlist_QP36.m3u8
#EXT-X-STREAM-INF:PROGRAM-ID=1,BANDWIDTH=505000,
http://example.com/playlist_QP40.m3u8
#EXT-X-STREAM-INF:PROGRAM-ID=1,BANDWIDTH=360000,
http://example.com/playlist_QP44.m3u8
#EXT-X-STREAM-INF:PROGRAM-ID=1,BANDWIDTH=277000,
http://example.com/playlist_QP48.m3u8
```

Figure 8.17   Super playlist file for an eight-level fragmented video.

```
#EXTM3U
#EXT-X-KEY:METHOD=NONE
#EXT-X-TARGETDURATION:10
#EXT-X-MEDIA-SEQUENCE:94
#EXT-X-KEY:METHOD=NONE
#EXTINF:10,101.ts
101.ts
#EXTINF:10,102.ts
102.ts
#EXTINF:10,103.ts
103.ts
```

Figure 8.18   Dynamic playlist file for 10s chunks.

8.3.4   Media Control

HLS can be configured as a single low bit rate 64K audio stream. If the bandwidth of the lowest video bit rate cannot be supported, the device will fall back to this audio-only option until the bandwidth increases and video playback can resume. The audio fallback is a better alternative than both video and audio media freezing or stuttering during a temporary bandwidth outage or bottleneck. Multiple audio bit rates can also be activated within a single stream. This configuration allows the audio quality to match up with the video resolution and bit rate. For instance, a 480p video can be paired with 64K audio, a 720p video can be paired with 128K audio, and a 1080p video stream can be paired with Dolby Digital 5.1 (AC3) audio, the standard for home theater, broadcast, and cinema surround sound. In addition, support for HE-AAC and HE-AACv2 provides the best possible stereo audio quality (up to 48 kHz) at low bit rates with good dynamic range.

Just like audio, multiple video resolutions can be configured in HLS to allow playback on multiscreen devices such as smartphones, tablets, and HDTVs, including different generations of the same device type. For example, the iPhone 3GS and iPod Touch support the Baseline H.264 profile (up to level 3.0), the iPhone 4 and iPad 1 support the Main H.264 profile (level 3.1), and the iPhone 4S+ and iPad 2+ support the High H.264 profile. To this end, the configuration control allows multiple resolutions to be paired with multiple H.264 profiles and levels within a single HLS stream.

Alternate media allows a user to select a variant playlist that can override the main content within a single HLS stream (e.g., alternate video angle or audio/language track). The client device plays the override media (i.e., audio or video) and suppresses any media of the same type from the main presentation, if present. This flexibility allows multiple variants of the same media to be presented without storing/downloading duplicate media or re-encoding the original content.

8.4   HLS OVER 4G AND 802.11

Figure 8.19 shows an experimental setup to study the performance of HLS. We employ a 596-s 720p HD video, which is H.264 encoded at 5.147 Mbit/s. In addition, 48 kHz audio is AAC encoded at 448 kbit/s. Hence, the total bit rate of the unfragmented video plus audio is 5.595 Mbit/s. This video/audio is segmented and re-encoded into 10s H.264 chunks with three video quality levels (350, 800, 1500 kbit/s) and one audio level (128 kbit/s). Thus, the bit rates (video + audio) for the three quality levels are 478 kbit/s (350 + 128), 928 kbit/s (800 + 128), and 1628 kbit/s (1500 + 128). The fragmented video/audio is streamed from an HTTP server using a high-speed Internet connection to a 4G cellular network (path 1). There are two different service providers for path 1: one for wireline Internet and another for wireless 4G. The same fragmented video/audio is also streamed locally via an 802.11n connection (path 2). The video/audio playback is achieved using HTML5 on the iPhone. Although the quality level may change when the connection automatically switches between 802.11n and 4G, smooth transition with no interruption in the video/audio playback is observed on the iPhone.

8.4.1   Startup Delay

Figures 8.20 and 8.21 show the initial startup time (i.e., the delay associated with sending the request and receiving the video for decoding before playback starts) for 4G and 802.11n, respectively. The HTTP sequence number is used to identify each

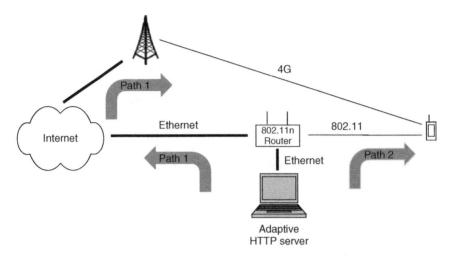

Figure 8.19   Experimental setup for HLS over 4G and 802.11n.

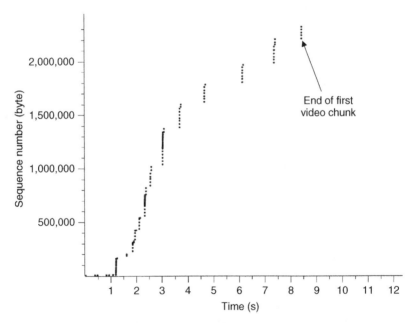

Figure 8.20   4G startup delay.

byte of data sent by the server. From the figures, the size of the first video chunk can be estimated to be about 2.4 Mbytes. The startup delay is 9s for 4G and 1.7s for 802.11n. Thus, the playback begins even if the entire 10s chunk is not received. The startup delay is dependent on the network delay and the number of quality levels. Employing more quality levels reduces the startup delay.

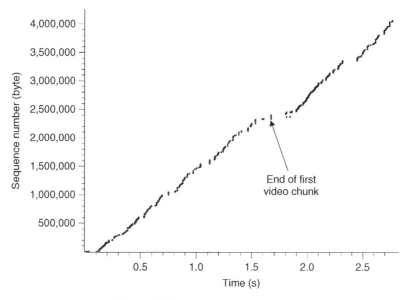

Figure 8.21   802.11n startup delay.

8.4.2   Switching Quality Levels

The quality levels may switch based on the iPhone's bandwidth estimation and memory resources. Figure 8.22 shows the HTTP GET requests sent by the iPhone to the video server when there are 802.11n and 4G connections. The super playlist (e.g., all.m3u8) is requested at the start of the streaming process whereas the dynamic

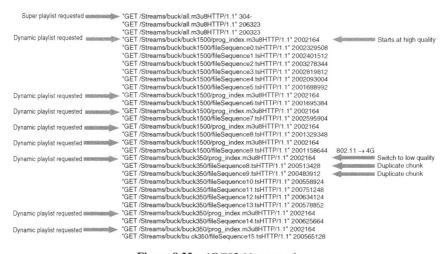

Figure 8.22   4G/802.11n streaming.

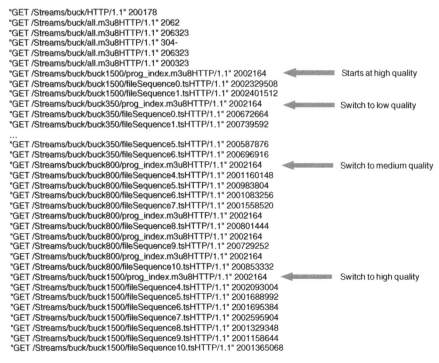

```
"GET /Streams/buck/HTTP/1.1" 200178
"GET /Streams/buck/all.m3u8HTTP/1.1" 2062
"GET /Streams/buck/all.m3u8HTTP/1.1" 206323
"GET /Streams/buck/all.m3u8HTTP/1.1" 304-
"GET /Streams/buck/all.m3u8HTTP/1.1" 206323
"GET /Streams/buck/all.m3u8HTTP/1.1" 200323
"GET /Streams/buck/buck1500/prog_index.m3u8HTTP/1.1" 2002164              Starts at high quality
"GET /Streams/buck/buck1500/fileSequence0.tsHTTP/1.1" 2002329508
"GET /Streams/buck/buck1500/fileSequence1.tsHTTP/1.1" 2002401512
"GET /Streams/buck/buck350/prog_index.m3u8HTTP/1.1" 2002164               Switch to low quality
"GET /Streams/buck/buck350/fileSequence0.tsHTTP/1.1" 200672664
"GET /Streams/buck/buck350/fileSequence1.tsHTTP/1.1" 200739592

...
"GET /Streams/buck/buck350/fileSequence5.tsHTTP/1.1" 200587876
"GET /Streams/buck/buck350/fileSequence6.tsHTTP/1.1" 200696916
"GET /Streams/buck/buck800/prog_index.m3u8HTTP/1.1" 2002164               Switch to medium quality
"GET /Streams/buck/buck800/fileSequence4.tsHTTP/1.1" 2001160148
"GET /Streams/buck/buck800/fileSequence5.tsHTTP/1.1" 200983804
"GET /Streams/buck/buck800/fileSequence6.tsHTTP/1.1" 2001083256
"GET /Streams/buck/buck800/fileSequence7.tsHTTP/1.1" 2001558520
"GET /Streams/buck/buck800/prog_index.m3u8HTTP/1.1" 2002164
"GET /Streams/buck/buck800/fileSequence8.tsHTTP/1.1" 200801444
"GET /Streams/buck/buck800/prog_index.m3u8HTTP/1.1" 2002164
"GET /Streams/buck/buck800/fileSequence9.tsHTTP/1.1" 200729252
"GET /Streams/buck/buck800/prog_index.m3u8HTTP/1.1" 2002164
"GET /Streams/buck/buck800/fileSequence10.tsHTTP/1.1" 200853332
"GET /Streams/buck/buck1500/prog_index.m3u8HTTP/1.1" 2002164              Switch to high quality
"GET /Streams/buck/buck1500/fileSequence4.tsHTTP/1.1" 2002093004
"GET /Streams/buck/buck1500/fileSequence5.tsHTTP/1.1" 2001688992
"GET /Streams/buck/buck1500/fileSequence6.tsHTTP/1.1" 2001695384
"GET /Streams/buck/buck1500/fileSequence7.tsHTTP/1.1" 2002595904
"GET /Streams/buck/buck1500/fileSequence8.tsHTTP/1.1" 2001329348
"GET /Streams/buck/buck1500/fileSequence9.tsHTTP/1.1" 2001158644
"GET /Streams/buck/buck1500/fileSequence10.tsHTTP/1.1" 2001365068
```

Figure 8.23   802.11n streaming.

playlist (e.g., prog_index.m3u8) is requested periodically. Each playlist or video chunk requires a separate GET request. Figure 8.23 shows the GET requests when streaming the video over a local 802.11n connection (no 4G connection), which has a maximum bit rate of 144 Mbit/s. In this case, the quality may transition from high to low and then back to high again when the sufficient bandwidth is available to support the highest quality. This suggests that some of these switchovers may be overly sensitive.

Duplicate chunks, such as chunks 8 and 9 in Figure 8.22, are requested by the iPhone and sent by the server when a change in quality level is required. These chunks contain the same content as the older chunks but are encoded with a different quality. They ensure a smooth transition and a faster switchover to the new quality. Fewer duplicate chunks tend to be requested when the quality switches from high to low (Figure 8.22) unlike the case when the quality goes from low to medium and medium to high (Figure 8.23). Thus, changing to a higher quality level may lead to higher network bandwidth consumption due to the larger number of duplicate chunks and the higher bit rate needed to send the higher quality chunks. Although unnecessary, this may be justified since bandwidth or device resources are abundant when a change to a higher quality is requested. Conversely, fewer duplicate chunks of lower quality should be requested when bandwidth or device resources are low. The older chunks received previously may be used for playback if the duplicate chunks arrive late. Note that both low- and high-quality chunks require the same amount of playback time because the chunk duration is fixed. Clearly, a substantial amount of video data (corresponding to the duplicate chunks) may be discarded due to changes in the quality level.

8.4.3   One-Level Versus Unfragmented HLS

Figure 8.24 shows the variation in the rates when one-level 10s video chunks and the unfragmented video are streamed using HLS over 4G cellular (no 802.11n connection). Video playback is smooth for both cases. With one-level fragmentation, quality switching is effectively disabled. Recall that the overheads for one-level fragmentation are comparable to the unfragmented video. A slowdown in the rates is possible when the video is fragmented into chunks. In this case, a 10s chunk is sent after 8–10s. This rate slowdown prevents the network and the iPhone from being overwhelmed with too much data. It also allows the HTTP streams to be scaled to serve many concurrent connections. Although it takes a longer time to receive the fragmented chunks compared to the unfragmented case, less buffering is required. In addition, a lower average rate is required for sending the 10s video chunks than the unfragmented video.

If a higher quality level is chosen, a longer time is required to initiate the slowdown due to the need to send a greater amount of video data at the beginning. For example, it takes about 35s to initiate the slowdown for the 1628 kbit/s level whereas only 10s is required for the 478 kbit/s level. The amount of time to initiate the slowdown is almost proportional to the quality level since 1628/478 × 10s or 34s. Moreover, a higher quality level demands a higher bit rate during the slowdown. For example, the slowdown bit rates are about 5–7 Mbit/s and 3 Mbit/s for the 1628 kbit/s and 928 kbit/s levels, respectively. Note that the 4G network bit rate closely matches the video bit rate during the chunk rate slowdown. For instance, the average 4G bit rate

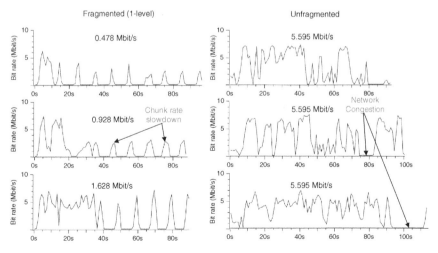

Figure 8.24   Fragmented and unfragmented 4G bit rate.

for the 478 kbit/s quality level is roughly $0.5 \times 3 \times 3$ or 4.5 Mbit/s. When averaged over a chunk slowdown period of about 10s, this gives 450 kbit/s.

In unfragmented video streaming, the video information is transmitted at the fastest rate allowed by the network. As a result, instances of network congestion are sometimes experienced. This is in contrast to the fragmented case, where self-imposed instances of nontransmission reduce the possibility of network congestion. It is interesting to note that there is no startup delay in the unfragmented video transmission. For the fragmented video, the initial handshake involving the exchange of super and dynamic playlists between the server and iPhone delays the start of the video transmission. This observation is consistent with Figure 8.20, which shows an initial handshake delay of just over 1s.

8.4.4 Multi-Level HLS

We employ the same experimental setup to stream the 720p HD video to the iPhone using multi-level HLS over separate 4G and 802.11n connections. The video is fragmented into 10s chunks with three quality levels that include audio. Figures 8.25 and 8.26 depict the respective 4G and 802.11n network throughput. The playback startup is faster than the one-level case because a match between the quality level and available network bandwidth can be obtained quickly for the multi-level case.

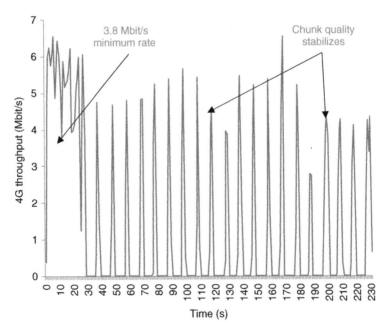

Figure 8.25   Multi-level HLS for 4G.

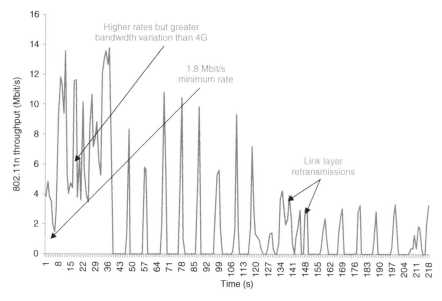

Figure 8.26   Multi-level HLS for 802.11n.

The bit rate variation at the beginning is caused by the iPhone testing the bandwidth of the connection to select the appropriate video quality level. The high initial rate allows the server to quickly send a large number of video chunks to the iPhone, which will buffer them before starting playback. The subsequent chunk rate slowdown reduces buffering requirements on the iPhone but there is no quality switching because the quality level has stabilized. Further observations can be made by comparing the slowdown regions of Figures 8.25 and 8.26. The 802.11n connection may require a greater amount of data buffering to cater for packet retransmissions at the link layer, which may be caused by radio interference. In contrast, 4G cellular operates on interference-free licensed spectrum and did not appear to suffer from such retransmissions.

Table 8.8 illustrates the network throughput (averaged over 135s) and the bit overheads for streaming the video using different quality levels. The average throughput is quite consistent for both 4G and 802.11n even though the instantaneous bit rates may exhibit higher variability. This demonstrates the robustness of HLS over two

Table 8.8   **Average Network Throughput and Relative Overheads (%).**

	478 kbit/s Level	928 kbit/s Level	1628 kbit/s Level
4G	729 kbit/s (52.5%)	1320 kbit/s (42.2%)	1977 kbit/s (21.4%)
802.11n	713 kbit/s (49.2%)	1240 kbit/s (33.6%)	2028 kbit/s (24.6%)

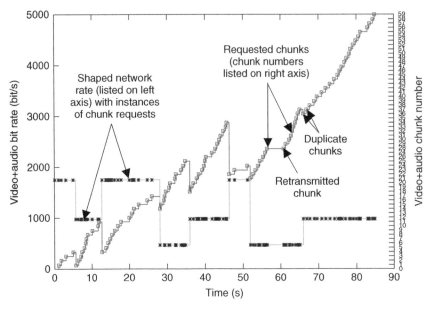

Figure 8.27   Quality switching with 2s video + audio chunks.

distinct networks: local-area 802.11n has far more bandwidth than wide-area 4G cellular but unlike 802.11n, 4G is not prone to radio interference. Clearly, the use of a lower quality stream increases the relative overheads.

8.4.5   Duplicate Video Chunks with Audio

Figures 8.27 and 8.28 illustrate the behavior of multi-level HLS over a dedicated 802.11n network using 2s and 5s video + audio chunks, respectively. The bit rates of the quality levels are 478 kbit/s, 928 kbit/s, and 1628 kbit/s. The actual network rates are higher since additional HTTP overheads are included. The server output rate is shaped to one of these rates for different time intervals. The quality level starts at 1628 kbit/s, dips to 928 kbit/s for several seconds (with a corresponding reduction in the video quality), and then returns to 1628 kbit/s and so on. The chunks are not received at fixed intervals even though the chunk duration is fixed. This is due to the delay jitter when the chunks are transmitted over a network.

When the quality level is changed, duplicate chunks are requested by the client and transmitted by the server. Hence more chunks are received. Typically, 3–4 duplicate chunks are requested on a switch from high to low quality whereas between 2 and as many as 13 chunks can be requested on a switch from low to high quality. When longer chunks are employed, the number of duplicate chunks tends to decrease. For example, 1 or 2 duplicate 5s chunks are requested on a switch from high to low quality. This is expected since longer chunks require a longer time for playback. Hence, fewer duplicate chunks are required. Again, a transition from low to high quality tends to

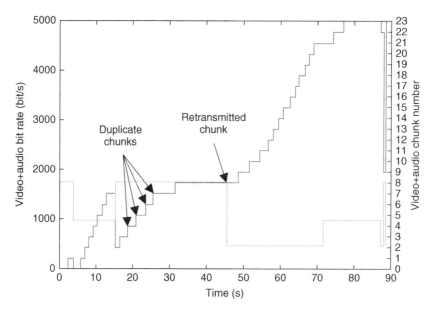

Figure 8.28 Quality switching with 5s video/audio chunks.

trigger a larger number of duplicate chunks than a transition from high to low quality. The last (penultimate) chunk at the old quality level is always requested again at the new quality level, which implies that a minimum of one duplicate chunk is needed when switching quality levels. Note that there is a distinction between duplicate and retransmitted chunks. Duplicate chunks are caused by changes in the quality levels whereas retransmitted chunks are caused by packet errors at the link layer.

8.4.6 Duplicate Video Chunks

Figures 8.29 and 8.30 illustrate the behavior of multi-level HLS over a dedicated 802.11n network using 2s and 5s video chunks, respectively. The bit rates for the quality levels are slightly lower (i.e., 350, 800, 1500 kbit/s) because no audio is included. For the 5s chunks, an extra quality level at 4500 kbit/s is used. Figure 8.29 clearly shows that more duplicate chunks are requested when the quality switches from low to high. For the period starting at 8s, only three duplicate chunks are requested when the quality switches from high to low at time 37s. For the period starting at 37s, 10 duplicate chunks are requested when the quality changes from high to low at time 68s. Since the elapsed time between quality switches are roughly the same (i.e., ~30s), the length of time between consecutive switches in quality may not be the cause for the larger number duplicate chunks when the quality level switches from low to high. It is interesting to observe that the period from 45 to 68s experiences a chunk rate slowdown where a chunk is sent after 2–3s. Such a slowdown may not occur if the quality levels switch frequently, as in Figure 8.27. The rate slowdown helps avoid network congestion.

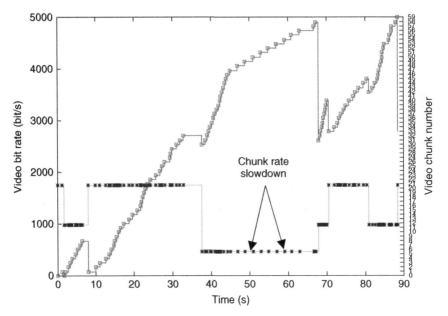

Figure 8.29   Quality switching with 2s video chunks.

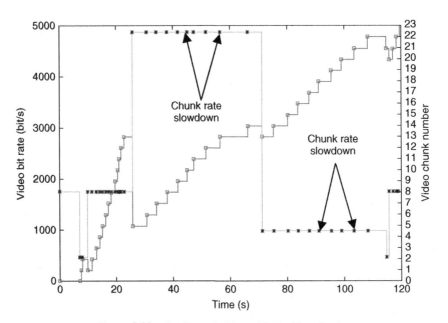

Figure 8.30   Quality switching with 5s video chunks.

As illustrated in Figure 8.30, more instances of chunk rate slowdown may occur for 5s chunks even though the slowdown bit rates may be higher than 2s chunks. In this case, a chunk is sent after 4 or 5s, which is longer than the slowdown interval for 2s chunks. Recall that the slowdown interval for 10s chunks is roughly 8–10s. Hence, the slowdown interval increases when longer chunks are employed. The difference in the quality levels during a switch, say from very high quality (e.g., 4500 kbit/s) to very low quality (i.e., 800 kbit/s) at time 70s, does not increase the number of duplicate chunks. In this case, only two duplicate chunks are requested. Note that the request for additional chunks is very responsive to rapid changes in the quality levels.

8.4.7 Duplicate Audio Chunks

In this case, 5s audio chunks with two quality levels at 128 kbit/s and 384 kbit/s are employed. As shown in Figure 8.31, the bit rate requirements are much lower because there is no video. Like the video chunks, more audio chunks are requested in a transition from low to high quality. Instances of chunk rate slowdown are also evident. The slowdown interval is about 9–10s, longer than the corresponding video chunks in Figure 8.30. This longer interval can be attributed to the lower bit rate of the audio chunks.

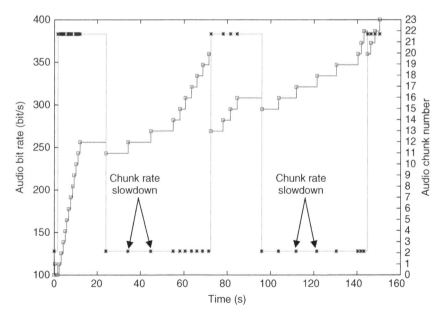

Figure 8.31   Quality switching with 5s audio chunks.

8.4.8　Duplicate Chunk Suppression

The impact of suppressing the duplicate chunks in HLS is assessed. Four-level video chunks with audio are employed. The bit rates (video + audio) for the four quality levels are 478 kbit/s (350 + 128), 992 kbit/s (800 + 192), 1756 kbit/s (1500 + 256), and 4884 kbit/s (4500 + 384). Thus, a higher quality video is accompanied by higher quality audio. The maximum amount of buffer space on the iPhone required to play the entire video is 4884 kbit/s × 596s or 363.9 Mbytes. The minimum amount of buffering is 478 kbit/s × 596s or 35.6 Mbytes, a 10-fold decrease from the previous case. The total buffer space needed at the server to store all four streams is (478 + 992 + 1756 + 4884) kbit/s × 596s or 604.2 Mbytes.

Duplicate chunks can be suppressed at the server (by ignoring the duplicate GET requests) or the iPhone (by not sending additional GET requests). However, even if the iPhone keeps the old chunks, the video playback stalls if the requested duplicate chunks are not sent by the server. This is because each video chunk may not start with an intracoded or key frame (i.e., open GOPs are employed). The stalling duration corresponds to the duration of the missing chunks. The audio is able to play without stalling because it is not constrained by the GOP structure.

8.4.9　Server-Based Chunk Suppression

One way to suppress the duplicate chunks without playback stalling is to allow the server to respond to the duplicate chunk requests sent by the iPhone. This can be achieved by keeping track of the last downloaded chunk (so that the server knows whether the iPhone is requesting for a duplicate chunk) and responding to every duplicate request by sending a 404 response (instead of video chunks) to the iPhone. In addition, the video chunks must be re-encoded and segmented to start with an intracoded key frame. Thus, all chunks can be synchronized and time aligned to start at the same time. Since the Apple video segmenter may not start each chunk with a key frame, an alternative segmenter that performs video transcoding by inserting a key frame at start of each chunk is used for this purpose. Table 8.9 shows that the chunk sizes using [2] and the original chunk sizes using the Apple segmenter are very similar (<0.3% difference). These chunks are encoded in the CBR mode.

Figures 8.32 and 8.33 show the impact of suppressing the duplicate chunks, which lead to fewer chunks received. This will not affect playback since the earlier received

Table 8.9　Re-encoded Chunk Sizes.

Quality Level (kbit/s)	Original Size (bytes)	Re-encoded Size (~5s/key frame)	Change in Chunk Size (%)
478	9,177,220	9,151,276	−0.28
992	16,283,620	16,288,320	0.29
1756	28,709,104	28,739,184	0.1
4884	77,925,436	78,110,804	0.24

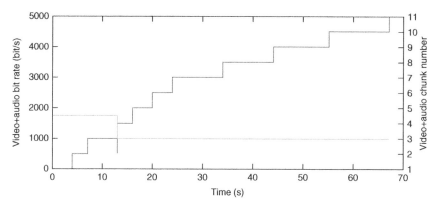

Figure 8.32   Duplicate chunk suppression by server.

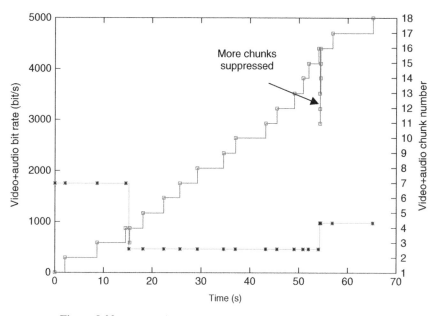

Figure 8.33   Aggressive duplicate chunk suppression by the server.

chunks can still be used for playback if the duplicate chunks did not arrive. More chunks are suppressed as the quality level switches from low to high.

Since more high-quality chunks are suppressed as the quality level switches from low to high, this leads to substantial bandwidth savings (in absolute terms). The converse applies when the quality level switches from high to low. As shown in Table 8.10, more than 80% relative bandwidth savings can be achieved. If the number of suppressed chunks exceeds the number of processed chunks (i.e., the old chunks), the relative bandwidth savings become more than 100%. Clearly, the relative bandwidth

Table 8.10   Bandwidth Savings with Duplicate Chunk Suppression.

Total Suppressed Chunks (bytes)	Total Processed Chunks (bytes)	Total Suppressed/ Total Processed (%)
8,935,640	10,524,052	84.9
5,014,900	12,579,456	39.9
10,836,884	21,340,632	50.8
9,976,972	12,979,708	76.9
742,224	15,271,804	4.9
1,244,748	23,721,840	5.2
7,998,648	14,248,332	56.1

savings is dependent on the number of suppressed duplicate chunks (at the new quality level) and the number of old chunks that are played back after a quality switch is made.

As an example, suppose the 478 kbit/s quality level is employed to stream ten 5s chunks for a 50-s video playback. The total amount of video data is $478 \times 10 \times 5$ or 23,900 kbits. If the 4884 kbit/s quality is chosen, the total amount of video data becomes $4884 \times 10 \times 5$ or 244,200 kbits. Suppose a switch to the 4884 kbit/s quality level is made after 25s of playback. Hence, five new higher quality chunks are needed to complete the remaining 25s of playback. If five duplicate chunks at the higher quality are suppressed during the switch, the absolute bandwidth savings become 244,200/2 or 122,100 kbits. This is same amount of bits to play the five higher quality chunks. Thus, the relative bandwidth savings is 100%. If six or more duplicate chunks are suppressed, the bandwidth savings exceed 100%. Conversely, if fewer than four duplicate chunks are suppressed, the savings is less than 100%.

8.4.10   Custom App Chunk Suppression

Figure 8.34 shows the duplicate chunk suppression performed by a custom app on the iPhone. The app employs a custom bandwidth-efficient player and chunks with aligned key frames. Since there is no duplicate chunk, redundant GET requests are not sent by the iPhone when a change in quality level is required. In addition, unnecessary transitions in quality levels are avoided. Clearly, this method is more bandwidth efficient than the server-based duplicate chunk suppression.

8.5   IMPACT OF VARYING CHUNK DURATION

Tables 8.11–8.15 show the various overheads for streaming a 120-s video as the chunk duration is varied but the quality level is fixed at 1756 kbit/s. The server output comprises all transmitted CBR chunks plus network overheads such as Ethernet, HTTP/IP headers, and HTTP acknowledgments. The GET overhead includes

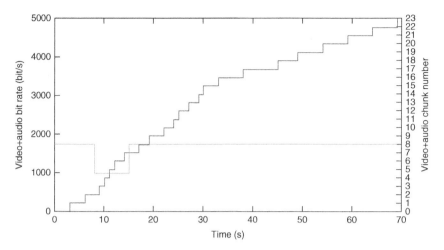

Figure 8.34 Duplicate chunk suppression using iPhone app.

Table 8.11   Overall Compressed Video File Size.

Chunk Duration (s)	Number of Chunks	Overall Video File Size (bytes)
2	60	26,950,552
4	30	26,933,820
5	24	26,930,812
10	12	26,923,480

Table 8.12   Playlist Overhead.

Chunk Duration (s)	Playlist File Size (bytes)	Number of Playlist Requests	Total Playlist Overhead (bytes)
2	2,157	3	6,471
4	1,136	12	13,632
5	920	13	11,960
10	515	10	5,150

the HTTP/IP headers. Table 8.11 shows that the overall compressed video file size decreases slightly as chunk duration increases from 2 to 10s. The playlist file size also decreases as the chunk duration increases (Table 8.12). As the chunk duration is reduced, more chunks are required, and hence the playlist file size and chunk overheads increase. Since a GET request is required to download each chunk from the server, more GET requests are needed for shorter chunks. Interestingly, shorter

Table 8.13   GET Overhead.

Chunk Duration (s)	Server Input (bytes)	Total GET Overhead (bytes)	Percentage of GET Overhead (bytes)
2	378,216	28,326	7.49
4	662.073	18,459	2.79
5	403,905	16,413	4.06
10	292,272	11,040	3.78

Table 8.14   Server Output Overhead.

Chunk Duration (s)	Server Output (bytes)	Output Overhead (bytes)	Percentage of Output Overhead
2	28,294,622	1,337,599	4.73
4	28,375,166	1,427,714	5.03
5	28,271,395	1,328,623	4.70
10	28,180,166	1,251,536	4.44

Table 8.15   Server Input and Output Transmission.

Chunk Duration (s)	Server Input/ Server Output (%)	Server Input + Output (bytes)
2	1.34	28,672,838
4	2.33	29,037,239
5	1.43	28,675,300
10	1.04	28,472,438

chunks tend to reduce the number of playlist requests. Hence, the 2s chunks lead to the lowest number of playlist requests. Recall that a GET request is also required to refresh a playlist. Since there are far more chunks than playlists (60 vs 3), the highest percentage of GET overhead correspond to the shortest chunk duration of 2s (Table 8.13). As can be seen from Table 8.14, the output overhead is fairly consistent, roughly 5% regardless of the chunk duration. The 10s chunks lead to the lowest ratio of server data input and output when compared to the other chunk durations (Table 8.15). This is expected since longer chunks minimizes the number of transmitted chunks and hence the bit overheads.

8.5.1   Impact of Varying Quality Levels

We now fix the chunk duration at 5s. The playlist file size becomes fixed at 920 bytes and the number of chunks stays at 24. The compressed file size increases as the

Table 8.16   Overall Compressed Video File Size.

Quality Level (kbit/s)	Number of Chunks	Overall Video File Size (bytes)
478	24	8,510,670
992	24	15,356,780
1,756	24	26,930,812
4,884	24	74,612,876

Table 8.17   Playlist Overhead.

Quality Level (kbit/s)	Playlist File Size (bytes)	Number of Playlist Requests	Total Playlist Overhead (bytes)
478	920	14	12,880
992	920	15	13,800
1,756	920	13	11,960
4,884	920	3	2,760

Table 8.18   GET Overhead.

Quality Level (kbit/s)	Server Input (bytes)	Total GET Overhead (bytes)	Percentage of GET Overhead (bytes)
478	121,517	16,679	13.7
992	369,390	17,450	4.72
1,756	403,905	16,413	4.06
4,884	491,110	13,346	2.72

quality level increases (Table 8.16). Higher quality levels tend to reduce the number of playlist requests (Table 8.17). Hence, the best quality level of 4884 kbit/s leads to the lowest number of playlist requests as well as the lowest absolute amount and percentage of GET overhead (Table 8.18). The output overhead stays at about 5% regardless of the quality level (Table 8.19). Thus, a higher quality level increases the output bandwidth consumption and the associated overhead quite linearly. The best quality level of 4884 kbit/s leads to the lowest ratio of server data input and output when compared to the other quality levels (Table 8.20).

8.5.2   Summary of HLS Performance

Table 8.21 summarizes the key metrics to minimize duplicate chunks using the native iPhone QuickTime player.

Table 8.19   Server Output Overhead.

Quality Level (kbit/s)	Server Output (bytes)	Output Overhead (bytes)	Percentage of Output Overhead
478	8,960,297	436,747	4.87
992	16,141,078	770,498	4.77
1,756	28,271,395	1,328,623	4.70
4,884	78,233,638	3,618,002	4.62

Table 8.20   Server Input and Output Transmission.

Quality Level (kbit/s)	Server Input/ Server Output (%)	Server Input + Output (bytes)
478	1.36	9,081,814
992	2.29	16,510,468
1,756	1.43	28,675,300
4,884	0.63	78,724,748

Table 8.21   Minimizing Duplicate Chunks and Streaming Bandwidth.

Metric	Impact
Chunk length	Longer chunks reduce the number of duplicate chunks
Number of quality levels	A smaller number of levels reduces the instances for quality change, which in turn reduces the number of duplicate chunks. However, this may be done at the expense of a longer startup delay and longer delay to switch between levels

Figures 8.35–8.38 summarize the playlist, GET, and input/output server overheads. We observe an identical correlation in the overheads for the 5s chunk duration and the 1756 kbit/s quality level. This is expected since the quality level stays at 1756 kbit/s when the chunk duration is varied whereas the chunk duration stays at 5s when the quality level is varied.

Figure 8.35 shows that the longest chunk duration (10s) or the best quality level (4884 kbit/s) will incur the least playlist overhead. Figure 8.36 is on a highly expanded scale on the vertical axis. It demonstrates that the server output overhead are roughly 5% of the total server data output and this is independent of the chunk duration and the quality level. Again, the longest chunk duration (10s) or the best quality level (4884 kbit/s) will lead to the lowest output overhead.

Figure 8.37 shows that the relative GET overhead can be minimized by choosing a 10s chunk duration or the best quality level of 4884 kbit/s. An interesting trend is observed in Figure 8.38 where the ratio of the server data input and output is closely

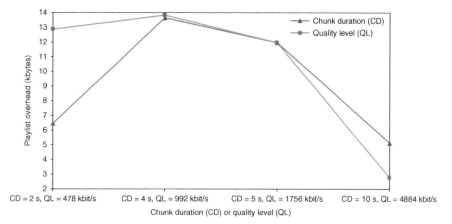

Figure 8.35 Impact of chunk duration and quality level on playlist overhead.

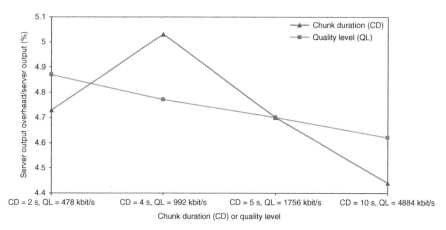

Figure 8.36 Impact of chunk duration and quality level on server output overhead.

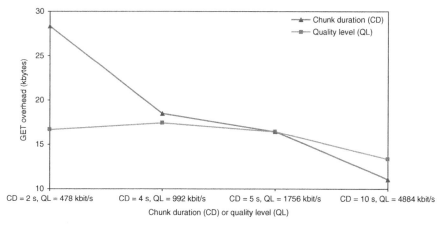

Figure 8.37 Impact of chunk duration and quality level on GET overhead.

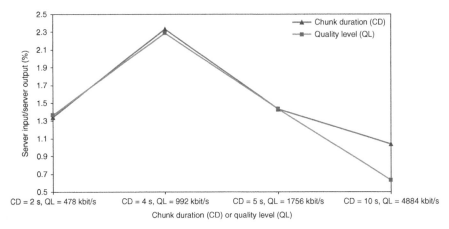

Figure 8.38 Impact of chunk duration and quality level on ratio of server data input and output.

matched for a broad range of chunk durations and quality levels. The highest ratio corresponds to the 4s chunks (with a quality level of 1756 kbit/s) and quality level of 992 kbit/s (with 5s chunks). The smallest ratio is achieved for 10s chunks (with a quality level of 1756 kbit/s) and a quality level of 4884 kbit/s (with 5s chunks).

Overall, the longest chunk duration or the best quality level (if bandwidth is available) should be employed whenever possible to achieve the best transmission efficiency in terms of the relative server input and output overheads. The chunk duration assumes higher priority over the quality level because the longest duration leads to the least absolute server data input and output (Tables 8.13 and 8.14) whereas the best quality level gives the highest absolute server data input and output (Tables 8.18 and 8.19).

8.6 MICROSOFT SILVERLIGHT SMOOTH STREAMING

MSS is based on the IIS media services extension that enables adaptive streaming of media to clients over HTTP. More specifically, it supports Windows Media HTTP, an HTTP-delivery platform from Windows Media Services (WMSP). WMSP uses standard HTTP for transferring data where the server maintains synchronized session states of the client (e.g., pause, replay, stop), effectively turning it into a streaming protocol such as Real-Time Transport Streaming Protocol (RTSP). MSS is the second most popular adaptive streaming standard after HLS. It is primarily used together with the Microsoft PlayReady content protection, Silverlight players, and Xbox. MSS can also be enabled on Apple and Android devices with a native application.

8.6.1 Overview of MSS Operation

The server sends the data packets to the client at the bit rate of the encoded media. The server only sends enough packets to fill the client's buffer, which is typically between 1 and 10s (MSS default buffer length is 5s). Thus, if a streamed video is paused, video data will not continue to download. Instead of managing thousands of small file segments, MSS employs the MPEG-4 Part 14 (MP4) container and stores each chunk as an MP4 movie fragment within a single contiguous MP4 file for each quality level. Hence, an MSS video is recorded in full length as a single MP4 file (one file per quality level) but is streamed to the client as an ordered sequence of fragmented MP4 file chunks.

The chunks are forwarded by an encoder to a Microsoft IIS server, which aggregates them for each quality profile into an "ismv" file for video and an "isma" file for audio. The "ismv" file contains the complete video with the chunks. Each "ismv" file corresponds to a video encoded at a specific quality level. If a video is encoded with different rates, several "ismv" files are produced. Unlike the Apple fragmenter, playback of each chunk is not possible because the chunks are embedded within the "ismv" file. However, the complete "ismv" video can be played back using the Windows Media player. An aggregate file format is used to store all the chunks and extract them when a specific request is made. For instance, the "ism" file specifies the bit rate of the "ismv" file. The "ismc" file contains the number of chunks and the chunk duration is controlled by the key frame interval, which can vary from 1 to 100s. Unlike MPEG-2 TS, the audio and video information can be transported as separate chunks if desired and then combined by the player.

The file starts with file-level metadata ("moov") that describes the overall video and the bulk of the payload is contained in fragment boxes that carry fragment-level metadata ("moof") and media data ("mdat"). The file is terminated with an index box ("mfra") that allows easy and accurate seeking within the video. The IIS server also creates an XML manifest file that contains information about the available quality levels. The HLS playlist specifies URLs. However, the MSS manifest file contains information that allows the client to create a URL request based on timing information in the stream (Figure 8.39). For on-demand service, the manifest files contain timing and sequence information for all the chunks in the video. Because metadata is provided in every chunk (the current chunk holds timestamps of the next chunk or two), this allows the client to access subsequent chunks without a refreshed manifest file. Hence, the manifest file need not be updated frequently. This is in contrast to HLS where, as new chunks become available, the playlist is updated to reflect the latest available chunks.

8.6.2   MSS Streaming over 802.11n and 802.16

We evaluate the performance of MSS over 802.11n and 802.16 or WiMAX (Figure 8.40). The maximum wireless bit rate for 802.11n and 802.16 are 144 Mbit/s and 3 Mbit/s, respectively. Video playback is performed using the Silverlight player on a Windows laptop. A chunk interval of 2s is chosen in all the tests. Thus, there are 53 chunks for *300* movie trailer (720p, 24 Hz, 106s), which is VBR-encoded using VC-1. At the IIS server, rates are recorded at 1s intervals. The quality levels and bit rates are as follows:

```
<StreamIndex Type="video" QualityLevels="2"
TimeScale="10000000" Name="video" Chunks="14"
Url="QualityLevels({bitrate})/Fragments(video=
{start time})" MaxWidth="1280" Max-
Height="720" DisplayWidth="1280" Display-
Height="720">
<QualityLevel Index="0" Bitrate="750000" Codec-
PrivateData="00000001274D401F9A6281405FF2E022000
007D20001D4C1280000000128EE3880" MaxWidth="640"
MaxHeight="360" FourCC="H264" NALUnitLength-
Field="4"/>
<QualityLevel Index="1" Bitrate="3000000" Codec-
PrivateData="00000001274D40289A6280A00B760220000
07D20001D4C12800000000128EE3880"MaxWidth="1280"
MaxHeight="720" FourCC="H264" NALUnitLength-
Field="4"/>
<c t="2489302977667"/>
<c t="2489324332333"/>
<c t="2489345687000"/>
<c t="2489367041667"/>
<c t="2489580588333" d="21354667"/>
<StreamIndex Type="audio" QualityLevels="2"
TimeScale="10000000" Language="eng" Name=
"audio_eng" Chunks="14" Url="QualityLevels
({bitrate})/Fragments(audio_eng={start time})">
<QualityLevel Index="0" Bitrate="31466" Codec-
PrivateData="1190" SamplingRate="48000"
Channels="2" BitsPerSample="16" PacketSize="4"
AudioTag="255" FourCC="AACL"/>
<QualityLevel Index="1" Bitrate="31481" Codec-
PrivateData="1190" SamplingRate="48000"
Channels="2" BitsPerSample="16" PacketSize="4"
AudioTag="255" FourCC="AACL"/>
<c t="2489301415778"/>
<c t="2489322749111"/>
<c t="2489344082444"/>
<c t="2489365415778"/>
<c t="2489578962444" d="21333334"/>
```

Figure 8.39   Sample MSS manifest file.

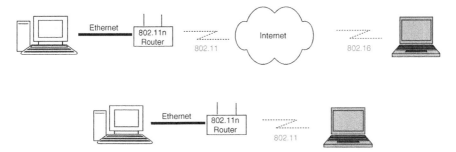

Figure 8.40   Experimental setups for MSS streaming.

- *Eight quality levels:* 230 kbit/s, 331 kbit/s, 477 kbit/s, 688 kbit/s, 991 kbit/s, 1.4 Mbit/s, 2.1 Mbit/s, and 3 Mbit/s;
- *Two quality levels:* 688 kbit/s and 331 kbit/s;
- *One quality level:* 3 Mbit/s or 688 kbit/s.

8.6.3   802.16 MSS Streaming

When the video is capped to a single quality level of 3 Mbit/s, the playback stalls constantly because 802.16 cannot handle the high video bit rate. Streaming stops at 375s, over three times greater than the video duration of 106s (Figure 8.41). When the video is to a single quality level of 688 kbit/s, playback stalling becomes less frequent because 802.16 can now handle the video bit rate although streaming still stops at 122s, 16s greater than the video duration (Figure 8.42). When the number of fragmentation levels is increased from 1 to 2, playback stalling also occurs less frequently and streaming stops at 120s, 14s greater than the video duration (Figure 8.43). As the

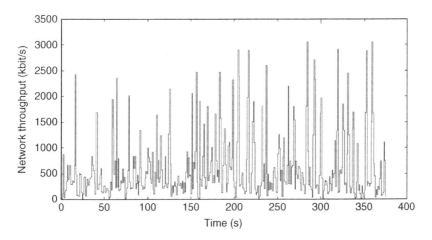

Figure 8.41   802.16 MSS (one-level *300* video capped to 3 Mbit/s).

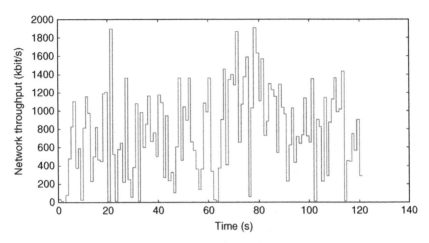

Figure 8.42 802.16 MSS (one-level *300* video capped to 688 kbit/s).

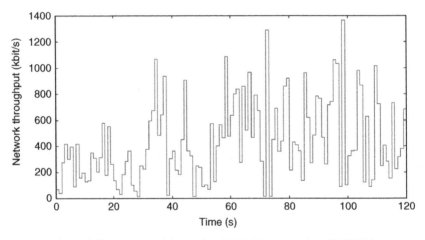

Figure 8.43 802.16 MSS (two-level *300* video capped to 688 kbit/s).

number of levels increases to 8 (Figure 8.44), playback becomes smooth but the video quality degrades frequently. In this case, frequent quality switching among the eight levels presents higher bit rate variability compared to one and two levels. Streaming stops at 77s, 29s less than video duration of 106s. Thus, the Silverlight player buffers about 29s of video.

8.6.4   802.11n MSS Streaming

In this case, playback is smooth and video quality is excellent because 802.11n can handle the highest video bit rate of 3 Mbit/s, even with only one quality level

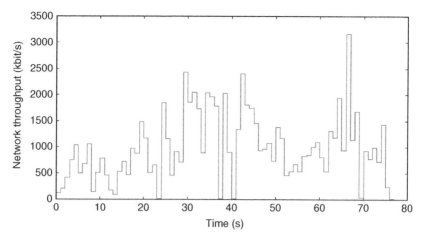

Figure 8.44   802.16 MSS streaming (eight-level *300* video capped to 3 Mbit/s).

(Figure 8.45). Unlike 802.16, fast start is evident here because of the higher 802.11n bandwidth in local area streaming. Although significant initial buffering is needed, video startup delay is reduced. The chunk request rate decreases as streaming progresses: one 2s chunk every 2s. The 802.11n network bit rate closely matches the video bit rate during the chunk rate slowdown. For instance, the average 802.11n bit rate is roughly 6 Mbit/s. When averaged over a chunk slowdown period of 2s, this gives 3 Mbit/s. For two-level fragmentation capped at 688 kbit/s (Figure 8.46), playback is also smooth. Streaming stops at 78s, 28s less than the video duration of 106s. For eight-level video chunks (Figure 8.47), playback is smooth and quality was excellent. However, higher bit rate variability is observed compared to one and

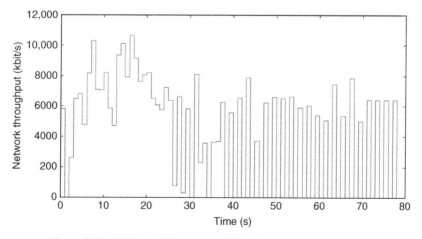

Figure 8.45   802.11n MSS (one-level *300* video capped to 3 Mbit/s).

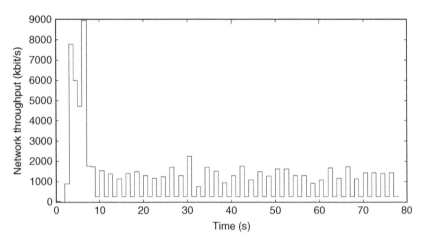

Figure 8.46   802.11n MSS (two-level *300* video capped to 688 kbit/s).

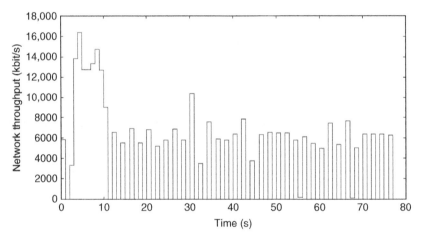

Figure 8.47   802.11n MSS (eight-level *300* video capped to 3 Mbit/s).

two levels. Streaming stops at 77s, 29s less than the video duration of 106s. A lower chunk request rate is observed as streaming progresses. It is interesting to note that the 802.11n network bit rate again matches the video bit rate during the chunk rate slowdown (i.e., 6 Mbit/s averaged over 2s). This suggests that even though eight quality levels are available, the higher 802.11n bit rates allow the best quality level of 3 Mbit/s to be chosen most of the time. This observation is further substantiated in Figure 8.46, which shows an 802.11n bit rate of about 700 kbit/s during the slowdown when two quality levels are available. Figures 8.48 and 8.49 show the higher chunk request rates at the start and the subsequent slowdown. Figure 8.48 also show that the manifest file is requested by the client at the start of the streaming but is not updated periodically.

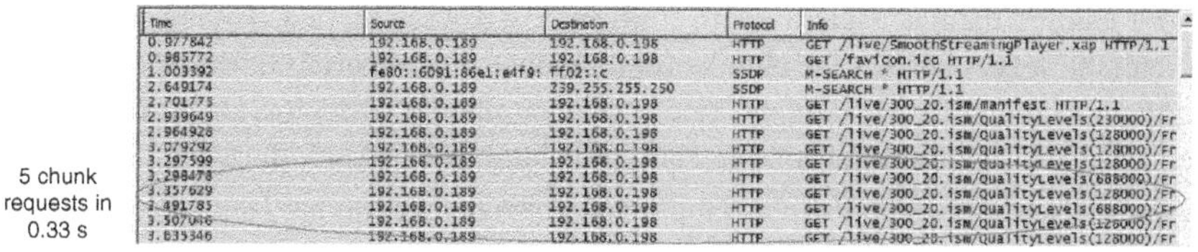

5 chunk
requests in
0.33 s

Figure 8.48 Higher chunk request rate at the start of streaming.

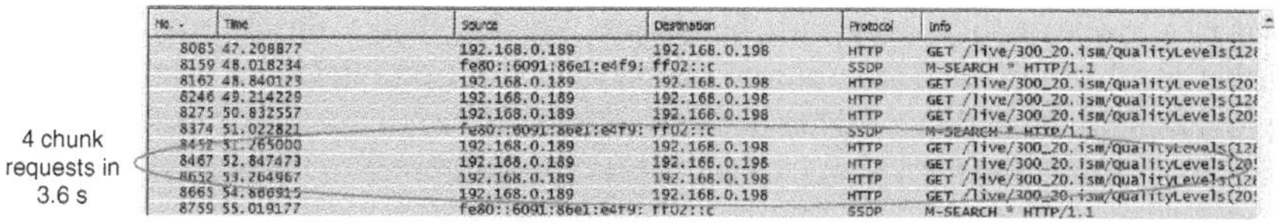

4 chunk
requests in
3.6 s

Figure 8.49 Lower chunk request rate at the later part of streaming.

Table 8.22 HLS and MSS 802.11n Streaming.

	MSS	HLS
Player startup delay	1–2s (independent of number of quality levels)	8–10s for one quality level, 1–2s for eight quality levels
Video data buffering	About 30s	Rate dependent (more buffering for higher rate)
Number of quality levels	More levels reduce player stalling but increase delay in quality switch	More levels reduce player stalling but increase delay in quality switch
Chunk rate slowdown	One chunk every few seconds	One chunk every few seconds
Duplicate chunk(s) on video quality switch	None	One or more chunks
Playlist or manifest file	Requested at start of streaming	Requested periodically

8.6.5 Comparison of HLS and MSS Streaming

The key differences are summarized in Table 8.22.

8.7 TRAFFIC RATE SHAPING

In an adaptive streaming system, if the server chooses to send the video data at a lower rate, the client will request for a lower quality level because it will erroneously "detect" a lower network bandwidth. The process of throttling the video data rate is called traffic rate shaping and this can be achieved by buffering the video chunks before sending them at a later time to conform to a desired bit rate. To shape the output bit rate, the server buffers the chunks for a predetermined period of time before sending them to the network. This bit rate can be computed using the ratio of the chunk file size (bits) and the total time to send the chunk (i.e., chunk duration plus buffering delay). For example, if the average size of each compressed frame is 100 kbytes for a specific quality level, then a 30 Hz video will require a minimum bit rate of 24 Mbit/s in order to play the video at the correct frame rate. This rate is independent of the chunk duration. However, if the server delays the sending of each 10s chunk by 2s, then the bit rate becomes $(100 \times 1000 \times 8 \times 30 \times 10)/(10 + 2)$ or 20 Mbit/s. Hence, a lower quality level must be chosen. For 2s chunks, the bit rate decreases further to 12 Mbit/s. The variation of the bit rates for server delays of 1–4s is shown in Figure 8.50. Note that a higher dynamic range between the highest and lowest bit rates can be obtained for a larger server delay. The difference in the peak rates for the 2s and 4s server delay is 1.81 Mbit/s, giving a relative difference of just 9%.

8.7.1   Impact of Shaping and Scene Complexity on Quality Switching

We evaluate the impact of traffic shaping on HLS over a short-range 802.11n wireless network. The experimental setup is shown in Figure 8.51. The video playback on the iPad tablet is performed using a QuickTime player on HTML5. The *300* movie trailer (720p, 24 Hz, 106s) is fragmented into 2s chunks and pre-encoded using H.264 in the VBR mode with eight quality levels ranging from QP 20 (high quality) to QP 48 (low quality) in steps of 4.

Figure 8.52 shows the case when the server bit rate is throttled to 0.5 Mbit/s for 25s, 4 Mbit/s for 20s, and 1 Mbit/s for 20s. Because the *300* trailer begins with several dark scenes of low complexity followed by scenes with higher complexity after 10s, a drastic switch in quality from QP 20 to QP 48 (i.e., very high to very low quality) was initiated when the iPad "detects" a low network bit rate. This phenomenon can

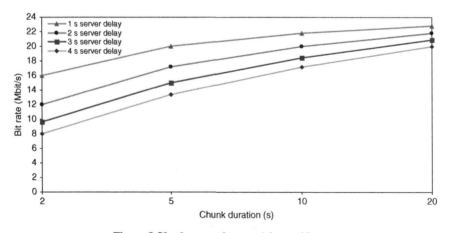

Figure 8.50   Impact of server delay on bit rate.

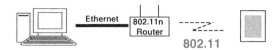

Figure 8.51   Experimental setup for HLS over 802.11n.

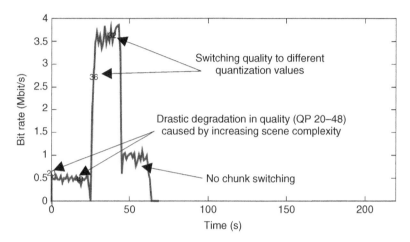

Figure 8.52   Quality switching with changes in QP values.

be explained as follows. The initial dark scenes require only low bit rates so even if the server output bit rate is shaped to a low rate, these scenes can be requested at the highest quality (i.e., QP 20). However, the subsequent high-complexity scenes require a very high bit rate for VBR-coded chunks, which cannot be met by the 0.5 Mbit/s rate. Thus, the lowest quality level (i.e., QP 48) is requested by the iPad. Note that if CBR-coded chunks are employed, the quality switch may not be required. However, the video quality may be affected.

From Figure 8.13, we note that the average chunk bit rate (QP 20) for the first 10s is low, which can be accommodated by the 0.5 Mbit/s rate. This triggers a sequence of five or more chunks at QP 20 from 0 to 10s. After 10s, the chunk bit rate increases sharply to 2.5 Mbit/s, which mandates a switch to a lower quality level (i.e., a higher QP). Figure 8.16 shows that for QP 40, the chunk bit rate is about 400 kbit/s from 10 to 12s. However, this rate is computed based on the encoded video and does not include the HTTP/IP overheads. In addition, the average chunk bit rate for QP 40 is nearly 1 Mbit/s from 24 to 26s. Hence, a sequence of eight or more chunks at an even higher QP value of 48 is chosen from 10 to 26s.

Since the network bandwidth has been held constant at various intervals, this suggests that content also plays a role in quality switching: when high-complexity scenes appear, lower quality chunks may be requested by the iPad to reduce the bit rate. The iPad's bandwidth estimator is responsive to a bit rate increase. As the bit rate rises to 3.5 Mbit/s, the quality level improves from QP 48 to QP 36 to QP 32. Because these

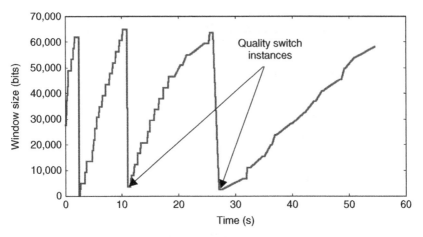

Figure 8.53   Advertised (receiver) window size during quality switch.

higher quality chunks are received at a very fast rate, there is no switch to a higher QP even when the bit rate of the server reduces later to 1 Mbit/s.

Tracking the advertized (receiver) window size may point to possible congestion and quality switching, as can be seen in Figure 8.53.

8.7.2   Impact of Shaping on Quality Switch Delay

The quality switch delay decreases when more bandwidth is available. As shown in Figure 8.54, the quality switch delay decreases as the shaping rate increases: 6s (0.5 Mbit/s), 4s (1 Mbit/s), 3s (1.5 Mbit/s), 2s (2.5 Mbit/s), and less than 1s (7.5 Mbit/s). When the shaping rate is 8 Mbit/s, no quality switching is needed. A similar trend is obtained for MSS using the same video with eight quality levels.

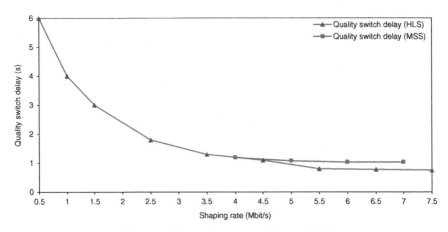

Figure 8.54   Quality switch delay for different shaping rates.

8.7.3   Impact of Shaping on Playback Duration

Figure 8.55 presents the playback duration for the *300* video of length 106s. The video is H.264-encoded using a QP value of 20 and the entire video has an average bit rate of 6.564 Mbit/s. Various chunk durations are employed and compared with the unfragmented case. Because only one quality level and high quality encoding are employed, the playback stalls quickly if a low shaping rate is employed. For example, for a 0.5 Mbit/s shaping rate, both fragmented and unfragmented playback stalls after 8s. This is expected since the shaping rate of 0.5 Mbit/s is much lower than the average chunk bit rate of 3.5 Mbit/s (10s chunk duration) and 2.5 Mbit/s (2s chunk duration) after 10s, as shown in Figures 8.12 and 8.13. However, as the shaping rate increases, the playback duration improves. For a shaping rate of 7.5 Mbit/s, the entire video can be played smoothly for chunks that are 5s or longer. Thus, shorter chunks (i.e., 2 and 4s) tend to suffer from longer playback stalling. The stalling duration is the longest for the unfragmented video, which shows the benefits of fragmentation, even for a single quality level. For a shaping rate of 8 Mbit/s, both fragmented and unfragmented videos can be played in full. Note that the peak chunk bit rates are 10 Mbits (10s chunks), 15 Mbit/s (2s chunks), and 30 Mbit/s (unfragmented).

Figure 8.56 presents the case for QP 40. Because the average rate for the entire video is now 0.517 Mbit/s, a shaping rate of 1 Mbit/s allows full video playback. The video chunks encoded with QP 40 reduce the stalling interval because less memory space is needed to store these lower quality chunks. A longer chunk further reduces the stalling interval. The results show that the appropriate shaping rate should be chosen depending on the quality of the encoding, the use of fragmentation, and the chunk duration.

8.7.4   Impact of Shaping on Start of Playback

Figure 8.57 shows the start of the playback for different chunk durations. Although the shaping rates may range from 0.5 to 8 Mbit/s, the same curve is obtained. When

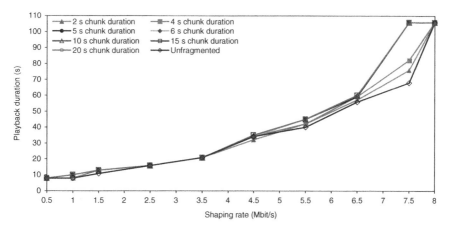

Figure 8.55   Playback duration for *300* video (QP 20, 106s).

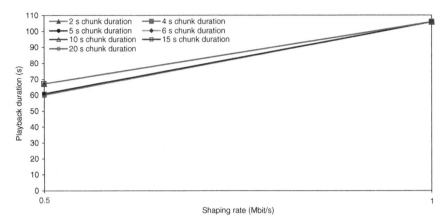

Figure 8.56 Playback duration for *300* video (QP 40, 106s).

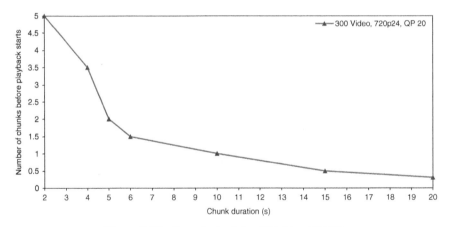

Figure 8.57 Start playback for *300* video (QP 20).

QP 40 (instead of QP 20) is employed, again the same curve is obtained. For 2s chunks, five chunks must be completely received before playback starts. For 10s chunks, one complete chunk must arrive before playback starts, otherwise the player will stall even as subsequent chunks are received later. Chunks can be received quickly if a higher shaping rate is used. Thus, the start playback will be faster for a shaping rate of 8 Mbit/s than 1 Mbit/s even though all chunks are of the same duration and require the same amount of playback time regardless of quality level.

8.7.5 Impact of Shaping and Scene Complexity on Duplicate Chunks

Figure 8.58 shows the case when the server output bit rate was initially shaped to 1 Mbit/s. After 60s, shaping is removed and the peak rate may reach over 20 Mbit/s.

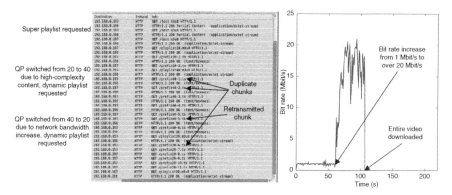

Figure 8.58 Duplicate chunks for *300* video.

When high-complexity content starts to appear after 10s, the quality switches to a lower level. Two duplicate chunks were requested. However, with the bit rate increase, only one duplicate chunk was requested. This phenomenon differs from the iPhone, where more chunks are requested on a switch to a higher quality level. One chunk was retransmitted due to an error in the original transmission.

8.7.6 Impact of Unshaped Traffic on Quality Switching

Figure 8.59 illustrates the case when the bit rates are not shaped. Because high bit rates are available and the iPad has adequate memory resources to store many chunks in advance, quality switching is not required and the best quality level at QP 20 is selected. Thus, even as the rates vary rapidly and the scenes of high complexity start to appear, many chunks can be buffered, a consistent quality level can be maintained,

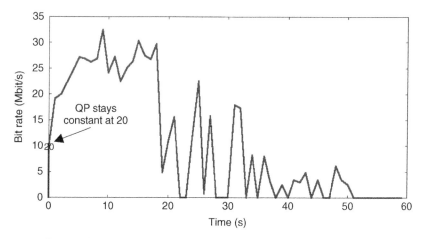

Figure 8.59 No quality switching when rates are unshaped (*300* video).

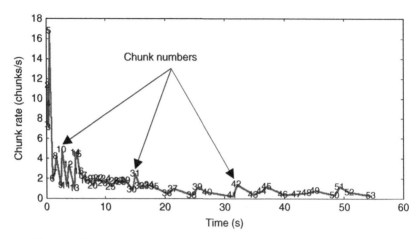

Figure 8.60 Chunk rate when quality switching is not required (*300* video).

and delays caused by quality switching can be avoided. Recall that in Section 8.4.3, a slowdown in the bit rates reduces memory resources on the iPhone but there is no quality switching during the slowdown. Figure 8.60 shows a high initial chunk rate but the rate decreases with time. Note that the entire video is received in just over 50s, which is less than half the length of the video.

8.8   ADOBE HTTP DYNAMIC STREAMING

Adobe's HTTP dynamic streaming (HDS) inherits technologies from Flash and Real-Time Messaging Protocol (RTMP). The protocol provides multiplexing and packetization services for the higher-level multimedia stream protocol, ensuring timestamp-ordered end-to-end delivery of all messages when used in conjunction with HTTP (Figure 8.61). In addition to the play command, RTMP incorporates a play2 command that can switch to a different rate bitstream without changing the timeline of the content played (Figure 8.62). This is useful for implementing random access and trick modes.

8.9   MPEG-DASH (ISO/IEC 23009)

The dynamic adaptive streaming over HTTP (DASH) standard aims to define an inter-operable delivery format that provides end users with the best possible media experience by dynamically adapting the media to changing network conditions [3]. The yet-to-be-ratified standard supports on-demand, live, and time-shifted applications and ads can be inserted between segments. It specifies use of either fragmented MP4 or MPEG-2 TS chunks. The media presentation description (MPD) is an XML manifest that is repeatedly downloaded. It is a key element of the standard and describes

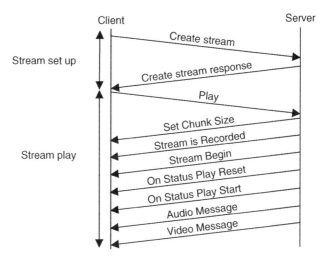

Figure 8.61   Stream setup and play in RTMP.

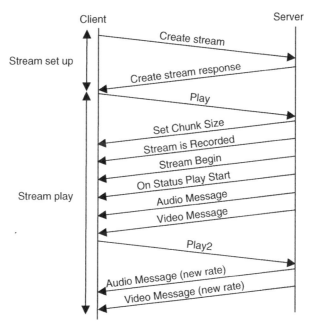

Figure 8.62   Switching streams in RTMP.

the accessible segments and corresponding timing. This enables quality or bitstream switching where video segments or chunks from different representations can be aligned and combined to a single conforming bitstream. Segment durations can be variable and the supported digital rights management (DRM) scheme can be signaled in the MPD. The standard specifies formats that:

- Enable provisioning and delivering media using existing HTTP-delivery networks and supports dynamic adaptation and seamless switching;
- Define the MPD containing XML-structured formats to announce Segment-URLs and to provide context of segments in the media presentation for selection and switching;
- Media segment formats containing efficiently coded media data and metadata according to or aligned with common media formats such as the ISO base media file format (BMFF) and MPEG-2 TS.

Part 1 of the standard was published on April 2012 and focuses on MPD and segment formats. Corrigenda and amendments for improved live metadata were subsequently issued. Parts 2 and 3 of the standard are in progress. Part 2 provides the conformance and reference software whereas part 3 describes the implementation guidelines.

8.9.1 DASH Process

The client downloads the MPD file and then downloads chunk by chunk based on the playout process. The bit rate is determined by the client. The Coordinated Universal Time (CUT) can be added in each segment to allow the client to control its clock drift for live sessions. Factors relevant for representation selection include the following:

- Buffer conditions;
- Network conditions;
- User changes resolution (e.g., from small to full screen);
- Device activity and resources.

8.9.2 DASH Media Formats

DASH supports two types of media segment formats, namely BMFF and MPEG-2 TS. If DASH is applied to H.264 videos based on BMFF, the H.264 file format is used for encapsulation of the video streams. Similarly, if DASH is applied to H.265 videos based on BMFF, the H.265 file format is used. One drawback of the H.264 file format relates to DASH-based live streaming of BMFF encapsulated H.264 videos. Due to a lack of parameter set track implementations of the H.264 file format, storing of parameter sets in sample entries becomes the only option in many cases. However, since sample entries must be included in the movie ("moov") box per BMFF, all parameter sets must be composed at the beginning of the streaming session and no new parameter sets may be generated afterwards. This may degrade the coding efficiency of the videos during live streaming. With the flexible storage of parameter sets in the H.265 file format, new parameter sets may be generated as needed and stored in-band with the video samples. DASH also supports carrying H.265 content that is packetized using the MPEG-2 TS format.

8.9.3 DASH for HTML5

This is under development via media source extensions. It allows the Java script to dynamically construct media streams for audio and video. The objectives are to:

- Allow a Java script to construct media streams independent of how media is fetched.
- Define a splicing and buffering model that facilitates use cases like adaptive streaming, ad-insertion, time-shifting, and video editing.
- Minimize the need for media parsing in Java script.
- Leverage the browser cache as much as possible.
- Provide byte-stream definitions for WebM and the ISO BMFF.

Google Chrome supports DASH based on WebM. Support for BMFF and more browsers are expected in near future.

8.9.4 DASH Industry Forum

The forum was initiated by the DASH Promoters Group in September 2012 [4]. It currently comprises 46 members; 12 members showed DASH demos at IBC 2012. The objectives of the forum are to:

- Promote and catalyze market adoption of MPEG-DASH;
- Publish interoperability and deployment guidelines;
- Facilitate interoperability tests;
- Collaborate with standard bodies and industry consortia in aligning ongoing DASH standards development and the use of common profiles across industry organizations.

8.10 AGGREGATE ADAPTIVE STREAM BANDWIDTH PREDICTION

An accurate prediction of the aggregate adaptive stream bandwidth from different transmitting sources is useful for bandwidth dimensioning in streaming or storage. Suppose that we wish to predict the total bandwidth of N video streams with M quality levels. Let $n_{k,k+1}$ be the number of quality changes between streams k and $k + 1$. The probability of transmitting N streams, each at quality level of λ_{α_i} where $\alpha_i \in \{1, 2, \ldots, M\}$ can be written as in (8.1):

$$p(\lambda_{\alpha_N}, \lambda_{\alpha_{N-1}}, \ldots, \lambda_{\alpha_1}) = \prod_{k=1}^{N} p(n_{k,k+1}) \quad (8.1)$$

The probability that the aggregate bandwidth is greater than Λ_s is shown in (8.2):

$$p(\Lambda > \Lambda_s) = \sum_{\alpha_1=1}^{M} \sum_{\alpha_2=1}^{M} \cdots \sum_{\alpha_N=1}^{M} p(\lambda_{\alpha_N}, \lambda_{\alpha_{N-1}}, \ldots, \lambda_{\alpha_1})$$

$$\text{where } \lambda_{\alpha_1} + \cdots + \lambda_{\alpha_N} > \Lambda_s \tag{8.2}$$

8.10.1 Permanence Time

In Chapter 2, we have demonstrated that the bit rate variation for videos of the same content is highly correlated even as the QP value is varied. This observation forms the basis of accurately predicting the aggregate adaptive stream bandwidth since adaptive streaming employs multiple streams of the same video that are encoded using different QP values (thus, giving rise to different quality levels). In addition to the QP value, the correlation also depends on start times of each stream and the permanence time at each quality level. The permanence time (T) refers to the time at a specific quality level and can be modeled as an inverse Gaussian distribution as in (8.3):

$$f(T) = \sqrt{\frac{\lambda}{2\pi T^3}} e^{\frac{-\lambda(T-\mu)^2}{2\mu^2 T}}$$

$$\text{where } \lambda = \text{Frame arrival rate}$$

$$\text{and } \mu = \text{Average frame duration} \tag{8.3}$$

For two concurrent streams, the start time (t) can be modeled as the sum of two independent exponential variables as in (8.4). This process can be extended to N streams:

$$f_1(t) = \mu_1 e^{-\mu_1 t} \quad f_2(t) = \mu_2 e^{-\mu_2 t} \tag{8.4}$$

8.10.2 Prediction Model Implementation

We illustrate the computation of the probabilities of the quality levels and permanence times using a single video stream. Table 8.23 shows a sample of the video trace employed. The trace can be segmented into chunks of about 5s long and the chunks are identified by 150 nominal quality levels. The probabilities of the different chunk bit rates (corresponding to different quality levels) for this trace is shown in Figure 8.63. Using this data, the probabilities of the different permanence times are obtained in Figure 8.64.

8.11 LIMITATIONS OF CLIENT-BASED ADAPTIVE STREAMING

Many adaptive streaming systems adopt a "client-pull" paradigm where the client device determines the quality level and the server accommodates the client's requests

Table 8.23   A Sample of the Video Trace.

Frame Number	Frame Type	Time (s)	Length (byte)
1	I	0	534
2	P	120	1542
3	B	40	134
4	B	80	390
5	P	240	765
6	B	160	407
7	B	200	504
8	P	360	903
9	B	280	421
10	B	320	461
11	I	480	1711

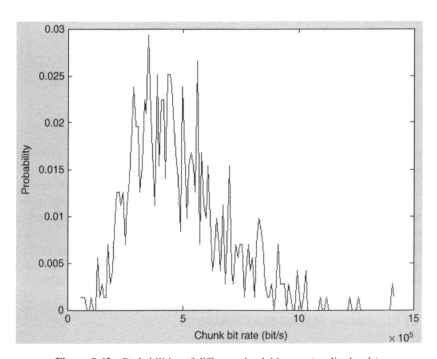

Figure 8.63   Probabilities of different chunk bit rates (quality levels).

by sending the appropriate video data. This requires a custom video player at the receiver (i.e., the client) to estimate the bandwidth. The playlist file at the server may also be requested periodically by the client so that the available video chunks can be selected based on the estimated bandwidth. If the bandwidth falls below a certain limit, chunks with higher quality levels (requiring higher bandwidth) will not

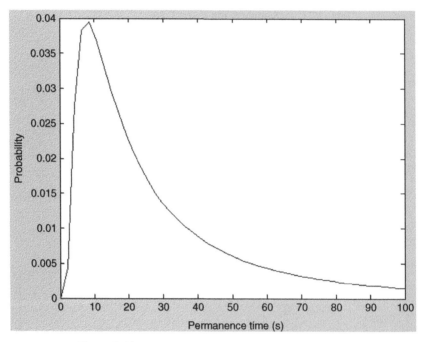

Figure 8.64   Probabilities of different permanence times.

be requested by the client. Depending on the number of quality levels, such a system may suffer from high startup delay and frequent quality switching, which leads to greater latency, especially with limited network bandwidth. In some instances, a drastic change in the video quality may occur from a severe fluctuation in the network bandwidth. The design of a bandwidth estimation algorithm with a proper switchover threshold is critical to the performance of client-based adaptive streaming systems but unfortunately the task is challenging. Some algorithms may constantly try to stream videos at the highest quality level. Others may choose to be conservative and stream at the lowest quality level when bandwidth for higher quality levels is available.

8.11.1   Limitations of Fixed-Size Chunks

It is interesting to note that in the late 1990s, there was a huge network paradigm shift from fixed-size asynchronous transfer mode (ATM) cells (which simplifies network switching fabrics and minimizes delay jitter) to variable-size IP packets (which minimizes bit overheads). It is natural to evaluate the possibility of using variable-size chunks in adaptive streaming systems. An obvious downside is that additional information on the number of frames contained in each chunk has to be relayed to the player at the client but this can be achieved by including a small amount of metadata at the start of each chunk. However, the bit rate of variable-size chunks is less susceptible to changes in the QP value caused by the video content. Thus, the suitable quality profile can be determined more accurately using the estimated network bandwidth. The file size (rather than the duration) for each chunk stays more or less constant. A chunk may contain fewer frames when specific sections of the video content contain high complexity or motion. Hence, the desired bandwidth for each chunk can be capped by the encoder without compromising video quality. In addition, a variety of encoding modes (e.g., strict VBR, capped-VBR, CBR) can be employed.

8.11.2 Server-Based Adaptive Streaming

An adaptive streaming system may allow the server to exert some control on quality switching to improve the adaptivity to network bandwidth variance and video content complexity. Such control is highly desirable for content or service providers because it allows the server traffic load to be managed, which reduces the possibility of a drastic change in the video quality that degrades end-user quality of experience. In the limiting case, bandwidth estimation and quality switching are performed entirely by the server, leading to a "server-push" paradigm. The number of quality levels that is selectable by the client can be regulated automatically by shaping the server's output transmission rate to match the bandwidth measured by the server as well as the average chunk bit rate. In this case, the adaptive streaming system is compatible with any legacy player because chunks with different quality can be concatenated at the server before they are sent out and no playlist is required. In addition, variable chunk durations can be used, delays due to chunk requests can be eliminated, and startup latency can be minimized. Although the current computing resources of the client device may not be available to the server, many tablets and smartphones are designed for HD video playback and should be capable of handling the quality of the chunks received from the server.

A server-based adaptive streaming system is shown in Figure 8.65. The source-controlled system allows partial or full control by the server in minimizing quality switch activity, thereby maintaining consistent video playback. The video encoder includes a fragmenter that may be integrated with the HTTP streaming server. The server includes a rate shaper and a bandwidth estimator. The encoder and server provide adaptive streaming services to the client device. The encoder determines the maximum number of quality levels based on a number of video encoding parameters such as the desired video resolution and bit rate. The encoder then compresses and fragments the video into chunks with different quality levels (possibly of the same chunk duration or of the same chunk file size). The number of levels and the bandwidth limit for each level are specified in the playlist file. These parameters may be modified dynamically by the server depending on variations in

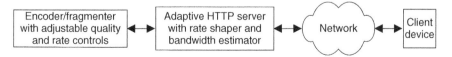

Figure 8.65 Server-based adaptive streaming system.

the network bandwidth or the video scene complexity, which can be detected by comparing the average chunk bit rate of successive chunks.

8.11.3   Linear Broadcast Systems

In traditional adaptive systems, quality switching and bandwidth estimation are performed by the client and chunks are played back in sequence at the device. In a server-based system, the server performs bandwidth estimation, quality switching, and if needed, concatenation of chunks with different quality. An unfragmented video bitstream or a regular fragmented MP4 or TS stream can be sent to the client, which need not issue any GET or playlist requests. Fragmentation overheads can be eliminated if the unfragmented bitstream is sent. There are several advantages when server-based adaptive streaming is extended to linear broadcast systems. In this case, bandwidth estimation is not needed since the channel bandwidth is fixed. For example, most digital TV and cable systems in the US employ fixed 6 MHz broadcast channels. Quality switching is based on video content complexity. In addition, a lower headroom on the overall network capacity is required when aggregating multiple video bitstreams. This headroom is dependent on the lowest quality level. No bandwidth feedback is required from the network or client.

8.11.4   Adaptive Streaming and Scalable Video Coding

Scalable video coding (SVC) requires a high-quality video bitstream that also contains one or more subset bitstreams. The subset bitstream is derived by the client device by dropping packets from the base video to reduce the spatial resolution, frame rate, or video quality. However, SVC incurs high processing overheads when adapting to dynamic changes in the bandwidth. High bandwidth is still needed to carry the high-quality bitstream. In contrast, adaptive streaming allows variable quality video chunks. The lowest quality is user-defined. Chunks can be independently decoded (i.e., each chunk can start with an intracoded frame). It is also faster to switch chunks than change video coding parameters.

8.12   TIPS FOR EFFICIENT ADAPTIVE STREAMING

We highlight some key deployment issues in the following section.

8.12.1   Quality Levels and Chunk Duration

More quality levels generally lead to more supported devices and a better user experience. Employing more quality levels also reduces instances of player stalling since it will make it easier for the client device to select the appropriate bandwidth level. However, this may increase the quality switch delay, which may become significant when network bandwidth is limited. In addition, more memory space is required to store the additional encoded bitstreams at the server. The frequency of quality switching should be minimized to maintain consistent performance. This is not only related to the number of quality levels and the network bandwidth variation but also the chunk duration. A greater number of levels can be selected if longer chunks are selected. Combining short chunks with a high number of quality levels may lead to frequent quality switching.

8.12.2   Encoder Efficiency

The maximum number of selectable quality levels is related to the encoder efficiency. Although four to eight quality levels seem optimal, two levels can be considered if the encoder is highly efficient. In the limiting case of one quality level, no quality switch delay is incurred and this may be useful for videos that are efficiently encoded for small screen displays. On the other hand, eight quality levels may be needed for transporting HD videos. In this case, since a higher number of quality levels leads to a higher probability of quality switching and higher bit rate variability, longer chunks may be desirable to reduce the switching frequency and overheads.

8.12.3   Bit Rates of Quality Levels

The bit rates for the quality levels should match the characteristics of the video (i.e., content, resolution, quality). A mismatch may lead to degradation in the video quality and viewing experience. The bit rates should also match the expected network bandwidth. A mismatch may disable quality switching altogether. This situation may arise if all quality levels other than the lowest quality level have a higher bit rate than the bandwidth. Such a situation may also arise when abundant bandwidth is available and the bit rate for the best quality level is well below that bandwidth. Due to the smaller bandwidth requirements, lower quality chunks are likely to arrive on time at the player. However, the video quality of these chunks should be assessed carefully since perceptible or visible degradation is likely to occur with lower quality levels.

8.12.4   Server Bandwidth Shaping

If multiple clients request for high-quality chunks simultaneously, this may induce a bandwidth bottleneck at the server as well as at the headend of the access network. Bit rate shaping (throttling) at the server helps alleviate this situation without additional hardware. By ensuring that the server has an independent mechanism for measuring the actual network bandwidth and shaping the output rate accordingly, unnecessary quality switching is avoided and the performance of the overall system is correspondingly improved. The rate shaper controls the number of quality levels by dynamically adjusting the server's output rate. For example, the number of quality levels can be reduced by shaping the output rate to a lower value. In this case, the higher quality levels (with higher bandwidth demands) are deactivated. In the limiting case, adaptive streaming can be disabled and the fragmented video is streamed using a single quality level. Among the benefits of performing rate shaping at the server include the following:

- Minimizing instances of oscillatory quality switching resulting from high network bandwidth variance;
- Reducing dependency on the client's player and the playlist for ensuring consistent video playback;
- Flexible multiscreen deployment where a broad range of quality levels can be set up but rates can be shaped to lower values for smaller screen displays;
- Improving the scalability of the traffic load on the server when servicing many concurrent sessions during peak periods.

Thus, by allowing the server to exert some control on the output rate, consistent video quality is maintained for the client device. Such control is particularly important for optimizing the performance of adaptive streaming over heterogeneous networks, including IP transport networks operating over wireless networks. Note that for a "server-push" adaptive streaming system, the server simply performs any necessary chunk switching after checking the available network bandwidth and average chunk rate. The client is more likely to estimate a bandwidth that is similar to the output rate shaped by the server.

8.12.5   Server Bandwidth Estimation

The server may measure the network bandwidth using a combination of metrics such as the advertised window of the receiving client, the congestion window of the server, and the IP packet retransmission statistics. The average chunk bit rate that relates to video complexity, resolution, and quality can be used to match the estimated bandwidth as well as buffering requirements for the specific chunk size.

8.12.6   Analyzing Network Congestion

The public Internet may be affected by congestion. The window size may change due to IP packet retransmissions and timeouts caused by congestion at the network layer. However, there is a need to decouple network and link/network layer interaction. For example, wireless networks are more error-prone than wireline networks. Packet retransmissions at the link layer are caused by bit errors at the physical layer (e.g., due to radio interference or weak signals). Such retransmissions are common for wireless networks but can be misinterpreted as congestion at the network layer (caused by packet timeouts), resulting in bandwidth throttling (caused by a reduction in the advertized window size) and an unnecessary reduction in video quality. Continuously sending data at a high rate over wireless tend to cause a higher probability of retransmission. Fortunately, many Wi-Fi routers can dynamically switch to noninterfering channels, forcing client devices to realign with channels specified by the router.

REFERENCES

1. HTTP Live Streaming, https://developer.apple.com/streaming, http://tools.ietf.org/html/draft-pantos-http-live-streaming-12.
2. Video segmenter, http://code.google.com/p/httpsegmenter.
3. MPEG-DASH, http://www.iso.org/iso/home/standards.htm.
4. DASH Industry Forum, http://dashif.org.

HOMEWORK PROBLEMS

8.1. HTTP adaptive streaming sends compressed video traffic by minimizing losses without necessarily increasing delay. Will HTTP adaptive streaming minimize delay in a lossless network? Can HTTP adaptive streaming operate properly when latency is not minimized?

8.2. Which metric assumes higher priority when a device decides to change the quality level (a) network bandwidth and (b) device memory resources?

8.3. How will the following metrics impact the startup playback delay (a) chunk duration, (b) network bandwidth, and (c) number of quality levels?

8.4. Like Transmission Control Protocol (TCP), HTTP employs a window mechanism to regulate the number of data packets that are sent over the network. The size of the congestion window dictates the number of outstanding packets without positive acknowledgment (ACK). For example, if the window size is k, the packet with sequence number n cannot be transmitted before the sender receives an ACK for packet $n - k$. In order to avoid congestion, the slow start mechanism is used to control the window size. The slow start mechanism is effectively a traffic shaping mechanism designed to throttle bursty data sources. Each negative acknowledgment (NACK) is assumed to be caused by packet losses due to network congestion and resets the window size to a small value (multiplicative decrease). The window will then restart using a slower additive increase process. The HTTP maximum segment length is 1448 bytes. HTTP flows normally start with an initial congestion window of at most three segments. Clearly, this may increase the delay to start the playback of a new video at the receiver when the video has just started streaming or when a retransmission has occurred. Estimate the startup delay for streaming 10s chunks of a compressed 25 Hz UHD video with an average frame size of 100 kbyte over a network with a bandwidth of 8 Mbit/s using HTTP. Will the playback start? What if 2s chunks are streamed instead? The chunk rate slowdown leads to bursty transmission at periodic intervals, which may trigger the slow start mechanism. Will it be better to transmit the chunks continuously at a lower rate?

8.5. The delay components for streaming live video in an end-to-end network are listed as follows:

- Encoding delay, E;
- Network delay (e.g., shaping, propagation, queuing), N;
- Resynchronization delay (e.g., assembling video frames from data packets and presenting them at a constant frame rate), R;
- Demultiplexing delay (if streams are statistically multiplexed), D.

The relation between these delay components and different network links is shown in Table 8.24. Give a comparative assessment of the relative values of the different delay components based on the video type. For instance, N may be larger for the Internet connection when compared to the dedicated connection, E may be larger for CBR than VBR encoding, E may be largest for adaptive streaming due to the need for segmentation, and so on. Is statistical multiplexing for CBR and adaptive streams useful? DTV and cable TV operators employ dedicated links to broadcast TV channels using IP packets. If video frames are directly transmitted instead of IP packets, will this remove the resynchronization delay, R? What are the drawbacks for doing this?

Table 8.24 Video Modes and Network Configurations.

Video Mode	Dedicated Link	Internet Link	Statistical Multiplexing
Raw	N	N, R	N, R, D
CBR	E, N, R	E, N, R	E, N, R
VBR	E, N, R	E, N, R	E, N, R, D
Adaptive	E, N, R	E, N, R	E, N, R

8.6. Is the cause for duplicate chunks due to the iPhone requesting for new chunks starting from the most recent chunk that begins with a key frame? Or is the iPhone requesting for new chunks starting from the chunk that was currently played?

8.7. In Table 8.9, the average size of a 5s chunk at 478 kbit/s is 478 kbit/s $\times$ 5 s or 2,390,000 bits. Hence, the number of chunks is $(9,177,220 \times 8)/(2,390,000)$ or about 30 chunks. Show that the number of chunks for the other three levels in the table is roughly 26 chunks.

8.8. Explain whether the following logic is sound. Over 100% relative bandwidth savings is likely to occur when the quality switches drastically from very low to very high. This is because a larger number of higher quality chunks are suppressed and the bandwidth requirements of these chunks are much higher than the low quality ones.

8.9. Explain whether the following logic is sound. The number of duplicate chunks requested may be dependent on the elapsed time since the last switch in quality level. A shorter elapsed time may lead to a larger number of duplicate chunks as many chunks that were requested previously by the iPhone may not have arrived. Since these higher quality chunks require larger bandwidth, short-duration quality switching to a higher quality should be avoided as much as possible.

8.10. Compare the advantages of using variable-size chunks when the chunks are encoded using VBR, capped-VBR, and CBR.

8.11. Consider an adaptive streaming algorithm that constantly tries to play the video at the highest quality level. How will this impact the chunk rate slowdown? Conversely, what is the impact on the chunk rate slowdown when a conservative algorithm constantly chooses the lowest quality level when bandwidth for the highest quality level is available? If the algorithm ensures fast startup and seek times by targeting the lowest bit rate before moving to a higher bit rate, how will this impact the chunk rate slowdown? For Figures 8.25 and 8.26, verify that the best quality level may not be chosen most of the time.

8.12. The first effort to adapt streams to client conditions was called stream thinning, which Microsoft introduced in 1998. When deteriorating network condition is detected, the video frame rate is reduced. In the worst case, the client may suspend video playback entirely. However, the connection is preserved and only audio is streamed. Compare stream thinning with adaptive streaming and provide a list of advantages and shortcomings for both systems. Will these systems work well in transcontinental networks with long signal propagation delays?

8.13. Unlike real-time streaming protocols designed for media delivery, the Internet was built on HTTP and optimized for nonreal-time Web data delivery. Web download services employ generic HTTP caches/proxies and are therefore less expensive than media streaming services offered by CDNs (with specialized streaming servers) and hosting providers. In addition, media protocols often experience difficulty in getting around firewalls and routers because they are commonly based on UDP sockets over unusual port numbers. HTTP media delivery uses the standard port 80 and does not require special proxies or caches. A media file is just like any other file to a Web cache. It is also much easier and cheaper to move HTTP data to the edge of the network, closer to users. Content delivery is dynamically adapted to the weakest link in the end-to-end delivery chain. Suppose the maximum size of each HTTP packet is 1448 bytes. For a video stream encoded at a quality level of 2 Mbit/s, calculate the number of packets to carry a 10s video chunk, assuming a 40-byte IP overhead for each packet.

8.14. For a 90-min sports program, compute the number of 2s chunks if four quality levels are required for MSS. Evaluate the size of the download buffer if the video has a frame rate of 30 Hz and is encoded at the highest quality level of 5 Mbit/s. Evaluate the pros and cons of using MSS's file management system where video chunks are stored within a full-length video. Will this allow the individual chunks to be decoded independently?

8.15. How will HLS and MSS respond to the following conditions?

- A change in the video scene complexity;
- There is sufficient network bandwidth but the user device does not have enough computing power to play the video at high quality;
- The video is paused or hidden in the background (e.g., minimized browser);
- The resolution of the best quality stream is larger than the screen resolution;
- A channel change;
- A change in the broadcast language;
- Insertion of multiple audio channels;
- Insertion of short advertisements/commercials.

8.16. Explain whether the chunk rate slowdown is suited for sports content with high motion.

8.17. Consider an integrated encoder and network streamer. In this case, a single video stream is required but the quality level can be varied dynamically depending on network conditions. There is no fixed duration for changing quality levels (i.e., the time granularity of a quality change is not limited by the chunk duration). Because the server need not maintain multiple streams of different quality for the same content, considerable storage space can be conserved. Thus, the least amount of advance video buffering is required and minimum delay is incurred when executing trick modes from the user device. Will more processing time be required for such a system?

8.18. Consider an adaptive streaming system that employs the display resolution or screen size to demarcate the different quality levels (e.g., 480p, 720p, 1080p). The same QP value is used for all resolutions. How does such a system compare with another system that uses a fixed resolution and only changes the QP value? If the following resolutions and rates are available in the super playlist, design a method to block a 720p HDTV from accessing the 320×180 resolution, which may result in perceptible video quality degradation.

Resolution	Video Bit Rate (Mbit/s)
1280×720	3
960×540	1.5
864×486	1.25
640×480	1
640×360	0.75
416×240	0.5
320×180	0.25

8.19. While traditional TCP only uses one connection path to send data, Multipath Transmission Control Protocol (MPTCP) can simultaneously use different connection paths, such as Wi-Fi and a smartphone's cellular connection, which may lead to better performance and resiliency. MPTCP has been used by some smartphone applications. Explain whether MPTCP is suitable for adaptive streaming.

8.20. Are there any advantages of performing bandwidth estimation at the server and the client?

8.21. Design an adaptive streaming system with multiple quality levels (say from the best to the worst levels) so that the activated level can be controlled by the server or the client for the following cases: the activated quality level corresponds to (a) the best level, (b) the worst level, and (c) any intermediate level between the best and worst levels.

8.22. Explain whether the following logic is sound. Longer chunks require fewer chunks to be transmitted, which in turn reduces the probability of a chunk error. Although more chunks are transmitted for shorter chunks, the bit error probability for shorter chunk is lower than a longer chunk (since there are fewer bits in a shorter chunk), which may reduce the number of chunk retransmissions.

8.23. An alternative approach to live or on-demand streaming to HLS-compatible players is to employ real-time transmultiplexing, which is offered by some streaming servers and CDNs. In this case, a video stream originally compatible with other formats (e.g., MSS, HDS) is dynamically converted to the HLS MPEG-2 TS chunks with the required manifest files. Discuss the trade-offs for doing this.

8.24. MPEG-DASH contains specific features that are lacking in HLS and MSS. They include definitions for the following:
- Segment format extensions beyond MPEG;
- Support for efficient trick mode;
- Simple splicing and (targeted) ad insertion;
- Definition of quality metrics;
- Profiles: restriction of features (claim, permission);
- Content descriptors (e.g., role, accessibility, rating, etc.).

Explain whether these enhancements are necessary.

8.25. Although not specified in HLS, consider the case when a playlist employs a cryptographic hash or checksum. By creating a cryptographic signature for the playlist, the entire content of the streams can be validated since any alteration to the file will invalidate the checksum. Explain whether this approach is better than directly encrypting the media content or using HTTPS authentication.

8.26. Consider an adaptive streaming system that employs longer chunks for lower bit rate streams. What are the key advantages of such a system compared to a traditional system that employs a fixed chunk size for all streams and bit rates?

8.27. HLS may produce hundreds or thousands of individual video chunks for each quality level. These chunks can be aggregated and then compressed using TAR or ZIP to reduce the number of files to be streamed. The compressed HLS files can be automatically decompressed for playback at the receiver. Explain whether this method will reduce the number of duplicate chunks.

8.28. What is the impact on the performance of adaptive streaming if the raw video is first compressed and then segmented into chunks versus the case when the raw video is first segmented and then each chunk is compressed individually?

8.29. Constant-duration chunks may lead to variable file sizes and hence higher delay jitter when the transmission bandwidth is fixed. An adaptive streaming system attempts to maintain a constant file size for every encoded chunk in order to conform to a specified rate cap. In other words, the rate cap is not used to cap the size of the encoded frame but is used to cap the chunk file size instead. To reduce the start-up delay, a small-big combo is employed for the first two chunks. The file size of first chunk is smaller than second chunk but is comparable to the third and subsequent chunks. The second chunk is the biggest chunk among all chunks and is roughly 1.5–2 times bigger. Since the first chunk will have started playback, the second chunk can be bigger to act as a buffer for subsequent chunks. These chunks require roughly the same transmission bandwidth (i.e., near CBR), which results in low delay jitter. Explain whether such a system is more suited for private managed networks or the public Internet.

GLOSSARY

1D	one dimensional
2D	two dimensional
3D	three dimensional
3GPP	Third Generation Partnership Project
4G	fourth generation
4K	four kilo
AAC	advanced audio coding
ABR	adaptive bit rate
ACK	acknowledgment
ADST	asymmetric discrete sine transform
ALF	adaptive loop filter
AMVP	advanced motion vector prediction
ASO	arbitrary slice ordering
ASP	advanced simple profile
ATSC	Advanced Television Systems Committee
AU	access unit
AVC	advanced video coding
AVI	audio video interleave

Next-Generation Video Coding and Streaming, First Edition. Benny Bing.
© 2015 John Wiley & Sons, Inc. Published 2015 by John Wiley & Sons, Inc.

BIFS	binary format for scenes
BLA	broken-link access
BMFF	base media file format
BO	band offset
CABAC	context adaptive binary arithmetic coding
CAVLC	context-adaptive variable-length coding
CATV	cable television
CB	coding block
CBR	constant bit rate
CDN	content delivery network
CIF	common intermediate format
COV	coefficient of variability
CPB	coded picture buffer
CRA	clean random access
CRT	cathode ray tube
CTB	coding tree block
CTU	coding tree unit
CU	coding unit
CUT	coordinated universal time
DASH	Dynamic Adaptive Streaming over Hyper-Text Transfer Protocol
DBF	deblocking filter
DCT	discrete cosine transform
DFT	discrete Fourier transform
DHCP	Dynamic Host Configuration Protocol
DIAL	discovery and launch
DLNA	digital living network alliance
DNS	Domain Name System
DP	data partitioning
DPB	decoded picture buffer
DRM	digital rights management
DST	discrete sine transform
DTV	digital television
DVB	digital video broadcast
DVD	digital video disk
DVI	digital visual interface
DVR	digital video recorder
EAS	emergency alert system
EBIF	enhanced television binary exchange format

EC	error concealment
EO	edge offset
FEC	forward error correction
FIR	Finite Impulse Response
FMO	flexible macroblock ordering
FRExt	Fidelity Range Extension
GDR	gradual decoding refresh
GOP	group of pictures
GPU	graphics processing unit
HD	high definition
HDMI	high definition multimedia interface
HDS	HTTP Dynamic Streaming
HE-AAC	high-efficiency advanced audio coding
HEVC	High Efficiency Video Coding
HLS	HTTP Live Streaming
HM	HEVC test model
HMAC	hash-based message authentication code
HRD	hypothetical reference decoder
HSS	hypothetical stream scheduler
HTTP	Hyper-Text Transfer Protocol
HVS	human visual system
Hz	short for frames per second
IDR	instantaneous decoding refresh
IEC	International Electrotechnical Commission
IIS	Internet Information Services
IP	Internet Protocol
IPCM	intra pulse code modulation
IPTV	Internet Protocol Television
ISO	International Standardization Organization
ISP	Internet service provider
ITU	International Telecommunications Union
ITU-T	ITU Telecommunication Standardization Sector
JM	joint model
JPEG	Joint Photographic Experts Group
JVT	joint video team
KLT	Karhunen–Loève transform
LCD	liquid crystal display
LED	light emitting diode

LM	linear model
MANE	media aware network element
MB	macroblock
MBAFF	macroblock adaptive frame-field
MBA	macroblock allocation
MDDT	mode-dependent directional transform
MIME	multipurpose Internet mail extension
MJPEG	Motion Joint Photographic Experts Group
MKV	matroska video
MOS	mean opinion score
MOV	movie
MP4	Moving Picture Experts Group-4
MPD	media presentation description
MPEG	Moving Picture Experts Group
MPTS	multi-program transport stream
MSE	mean squared error
MSS	microsoft smooth streaming
MV	motion vector
MVC	multiview video coding
NAL	network abstraction layer
NALU	network abstraction layer unit
NAT	network address translator
NRT	non real time
NSQT	non square quadtree transform
NTSC	National Television System Committee
OTT	over-the-top
PAR	peak-to-average ratio
PB	prediction block
PC	personal computer
Pel	pixel
PES	packetized elementary stream
PicAFF	picture adaptive frame-field
POC	picture order count
POP	point of presence
PPS	picture parameter set
PSNR	peak signal to noise ratio
PU	prediction unit
QCIF	quarter CIF

QoE	quality of experience
QoS	quality of service
QP	quantization parameter
RADL	random access decodable leading
RAP	random access point
RASL	random access skipped leading
RDO	rate distortion optimization
RDPCM	residual differential pulse code modulation
RGB	red–green–blue
ROI	region of interest
RPS	reference picture set
RQT	residual quad tree
RS	redundant slices
RTMP	Real-Time Messaging Protocol
RTP	Real-Time Transport Protocol
RTSP	Real-Time Transport Streaming Protocol
SAD	sum of absolute differences
SAP	sample-based angular intraprediction
SAO	sample adaptive offset
SATD	sum of absolute transform differences
SCC	scenarist closed caption
SD	standard definition
SDI	serial digital interface
SDN	software-defined network
SEI	supplementary enhancement information
SG	slice group
SI	switching intra
SMP	symmetric motion partitioning
SOP	structure of pictures
SP	switching predictive
SPS	sequence parameter set
SRT	SubRip
SSD	sum of squared differences
SSDP	Simple Service Discovery Protocol
SSID	service set identity
SSIM	structural similarity
STB	set-top box
STD	system target decoder

STSA	stepwise temporal sublayer access
SVC	scalable video coding
SVG	scalable vector graphics
SVOD	subscription video-on-demand
TB	transform block
TCP	Transmission Control Protocol
TMVP	temporal motion vector prediction
TS	transport stream or transform skip
TSA	temporal sublayer access
TSP	transport stream packet
TU	transform unit
UDP	User Datagram Protocol
UHD	ultra-high definition
UPnP	universal plug and play
URL	uniform resource locator
URQ	uniform reconstruction quantization
UVLC	universal variable length coding
VBR	variable bit rate
VBV	video buffer verifier
VCEG	Video Coding Experts Group
VCL	video coding layer
VHS	video home system
VLC	variable length coding
VOD	video-on-demand
VPS	video parameter set
VUI	video usability information
WEBVTT	Web video text tracks
WG	working group
Wi-Fi	wireless fidelity
WMSP	windows media services
WPP	wavefront parallel processing
Y-PSNR	luminance peak signal to noise ratio
YUV	luminance, red chrominance, blue chrominance

INDEX

Next-Generation Video Coding and Streaming, First Edition. Benny Bing.
© 2015 John Wiley & Sons, Inc. Published 2015 by John Wiley & Sons, Inc.

Printed and bound by CPI Group (UK) Ltd, Croydon, CR0 4YY

16/04/2025

14658363-0004